Molding For Profit:
The injection molding business in the 21st century

Clare Goldsberry

First published by Plastics Information Direct – a division
of Applied Market Information Ltd – in 2012
Republished by Elsevier

ELSEVIER

AMSTERDAM • BOSTON • HEIDELBERG • LONDON
NEW YORK • OXFORD • PARIS • SAN DIEGO
SAN FRANCISCO • SINGAPORE • SYDNEY • TOKYO

Molding For Profit: The injection molding business in the 21st century

First published by Plastics Information Direct – a division of Applied Market Information Ltd – in 2012
Republished by Elsevier

Elsevier
Radarweg 29, PO Box 211, 1000 AE Amsterdam, Netherlands
The Boulevard, Langford Lane, Kidlington, Oxford OX5 1GB, UK
50 Hampshire St, 5th Floor, Cambridge, MA 02139, USA

ISBN: 978-1-906479-07-7

Contents

Foreword .... **ix**

Chapter 1 Background – the evolution of the custom injection molding business **1**

1.1 Change in materials technology 2

1.2 Change in molding machinery technology 2

1.3 Change in products and markets 4

1.4 Change in captive and proprietary molding 7

1.5 Change in geographical locations 8

1.6 Globalization 9

1.7 Change in the manufacturing and business environment 11

Managing your business

Chapter 2 Developing marketing strategies for profitability **17**

2.1 What is marketing? 17

2.2 Creating a marketing strategy 18

2.3 Know your company 19

2.3.1 Know yourself – surveying employees 21

2.3.2 Know yourself – customer surveys 23

2.3.3 Using the data you collect 24

2.4 Know your customers 25

2.4.1 Using the data you collect 26

2.4.2 Identifying your value to your customers 27

2.4.3 Marketing based on value 28

2.4.4 New customers 31

2.5 Know your competition 31

2.6 Marketing methods 34

Appendix 2.1: Employee survey questionnaire template **36**

Appendix 2.2: Customer survey questionnaire template **38**

Chapter 3 Marketing communications – telling your story and communicating your value **40**

3.1 Branding isn't just for soft drinks 42

3.2 Communicating the right message 45

3.3 Websites 46
3.3.1 Priorities to show your audience 47
3.3.2 Internet marketing – more than just a website 50
3.4 Publicity tips: how to get noticed in the press 51
3.4.1 Trade publications 53
3.5 Trade shows and industry conferences 55

Chapter 4 Sales strategies 58

4.1 What do we mean by sales? 58
4.2 Why you need sales staff 59
4.2.1 Direct and indirect sales staff 60
4.2.2 Separating inside and outside sales 61
4.2.3 Who sells and what do they sell? 62
4.3 Why you need a sales strategy 63
4.4 Targeting your customers 66
4.4.1 Prioritise your sales efforts 68
4.4.2 When one customer equals too much business 69
4.4.3 New customers – same markets 70
4.4.4 New business from mold transfers 71
4.5 Internal communication is essential 74

Chapter 5 Business planning for success 76

5.1 Being successful in downtimes 76
5.2 Business planning 79
5.2.1 What is a business plan for? 79
5.2.2 Writing your business plan 80
5.3 Expanding a business through acquisition 84

Chapter 6 Employee training, hiring and firing 86

6.1 Training for success 86
6.2 Challenging times for employees 89
6.3 Challenging times for employers 90

Appendix 6.1: Skills audit and training plan 91

Understanding the industry

Chapter 7 Markets and profitability 97

7.1 Finding winning markets 97

7.1.1 Where to look for information on particular markets 99
7.2 Automotive/transportation 99
7.3 Medical devices/disposables/durables 105
7.3.1 European outlook for medical molding 109
7.4 Packaging – food, cosmetic, and consumer 111
7.5 Industrial 116
7.6 Aerospace/defense 117
7.7 Appliances 118
7.8 Toys, recreational and sporting goods 120
7.9 Building and construction 122
7.10 Electrical/electronic components 123
7.10.1 Automotive applications 123
7.10.2 Computer and connector applications 124
7.11 Consumer and personal care products 126
7.12 Summary 126

Chapter 8 Business challenges for the 21st century – local or global? 128

8.1 Finding your niche 128
8.1.1 Proprietary product niches for custom molders 130
8.2 Competing in a global market 131
8.2.1 Locating global facilities 132
8.2.2 Changes to the rules in Mexico 134
8.2.3 Offshoring and back-shoring 135

Chapter 9 Business challenges for the 21st century – sustainability 140

9.1 Lean manufacturing and sustainability 143
9.2 Energy and sustainability 145
9.3 Other sustainable ideas 146
9.4 Summary: areas to look for sustainability in your company 148

Chapter 10 Business operations for profitability 149

10.1 Learning from experience: what the observers saw in recent years ... 150
10.2 Lessons learned and progress toward profitability 152
10.3 Operations software 153
10.4 When big sales look tempting 156
10.5 Margins 157

10.6 Firing unprofitable customers 158

10.7 Typical signs of success 160

10.8 Profitability checklist 161

Technological challenges and opportunities

Chapter 11 Advanced injection molding technology – adding value and flexibility 165

11.1 Two-shot molding 165

11.2 Co-injection molding 168

11.3 In-mold decorating (IMD) and in-mold labeling (IML) 168

11.4 Film insert molding 170

11.5 Stack molding and tandem molding 171

11.6 Rotational stack molding 172

11.7 Structural foam molding 173

11.8 MuCell technology 174

11.9 Gas and fluid assisted injection molding 175

11.10 Micromolding 175

11.11 Choosing your process niche 177

Chapter 12 Factory organization and efficiency 178

12.1 More than molding machines 178

12.2 Robotics and molding manufacturing cells 179

12.2.1 Summary of potential benefits of incorporating robotics 181

12.3 Secondary operations 181

12.3.1 Welding 181

12.3.2 Decorating 182

12.4 Clean room manufacturing cells 183

12.5 Preventive maintenance saves money, extends equipment life 185

12.5.1 Precise wear measurement helps to avoid unscheduled downtimes 188

12.6 Summary 189

Chapter 13 Moldmaking technology 191

13.1 CAD/CAM software 192

13.2 Manufacturing cells for moldmaking 194

13.3 High-speed machining (HSM) 194

13.4 New technologies from moldmakers 195
13.5 Offshore mold manufacturing – to do it or not 196
13.6 10 Steps to being competitive in moldmaking 198
13.7 Make or buy mold components 199
13.7.1 The moldmakers' experience 199
13.7.2 The component supplier's viewpoint 201
13.8 Summary 202

Chapter 14 Additive manufacturing – from prototyping to production 203

14.1 Evolution of the business 203
14.1.1 New service bureau models 205
14.2 Latest materials and systems 206
14.2.1 Virtual prototyping 209
14.2.2 Direct metal laser sintering 209
14.3 Moldmakers and molders get in on the action 209
14.4 Making additive manufacturing a part of your business 212

Chapter 15 Polymer materials – minimize costs and maximize choice 216

15.1 The price of plastics 216
15.1.1 The price of scrap 217
15.2 Material developments 218
15.2.1 Automotive materials 218
15.2.2 Compounded materials for electrical and electronic applications 219
15.2.3 Materials in the personal care market 221
15.3 'Green' materials 221
15.3.1 The downside of bioplastics 223
15.3.2 What this means to molders 224

Final thoughts

Chapter 16 Planning for the longer term 229

16.1 What is the future of my business? Preparing to move on 229
16.1.1 Selling your business as a going concern 229
16.1.2 When is the right time to sell? 232
16.1.3 Partners and partnerships 233
16.1.4 Employee stock ownership plans 234
16.2 Summary 234

Additional material

Chapter 17 A case study **239**

Glossary of abbreviations **259**

Index **263**

Foreword

The custom injection molding business has never been easy. In comparison with some competing manufacturing industries, it's a relatively new technology that has evolved extremely quickly from the 1940s to the present day. And it's a technology which benefits from a diversity of raw materials and processing techniques, as well as being challenged by this same variety. However, perhaps the biggest challenges for those in the business of injection molding are the "business" issues.

Having spent 3 decades in and around this industry, I've experienced many of the problems described in this book, such as creating a brand, raising market awareness of a company when you don't have a recognised product to sell, and working out which customers are the "right" customers and how to sell to them. It's not easy and I've certainly not been alone in thinking this. There have been many company failures along the path. I have only to look back over my 20 year writing career in the industry to find those once thriving companies that I wrote about many years ago now vanished from the scene.

But in spite of these business challenges, many entrepreneurs over the past five decades have been extremely successful, and ultimately created some of the leading custom injection molding companies in the world today. They didn't do this by accident. They did it by planning, with purpose, and with the knowledge of how to create a successful business. If they didn't know all of this directly (and many didn't as they were either creative mold makers who wanted to get into molding, or technically savvy process engineers or technicians that wanted to work for themselves) they hired those who did know about business.

While I spent only about 10 years in the injection molding business itself – which was a valuable learning experience – much of what I know today has come from my 20 years of writing about the industry, interviewing hundreds of molders, moldmakers, materials specialists, money men and consultants who work within it. I've spent time picking the brains of engineers, purchasing agents and managers in large OEMs to try to understand their business strategies and how that can help the custom molders who supply them. These people have taught me just about all I know about creating a successful injection molding business.

If I tried to name everyone and thank them all individually, that would be a book in itself. Their words of wisdom appear in this book, taken from hours of interviews and literally hundreds of articles that have appeared in the plastics industry trade magazines for which I've been a contract writer.

I will thank two people who are no longer in the forefront of the industry publishing business: Suzy Witzler and Pete Sullivan, two of the people instrumental in starting Injection Molding Magazine. They gave me the initial opportunity to write my first book, *The Business of Injection Molding*, published by what was then Abby Communications, which also published my next four books. Because an update of all of this information was badly needed, AMI gave me this opportunity to publish **Molding for Profit: The injection molding business in the 21st century**.

So I will just say thank you to all of those people who, over my three decades in the injection molding industry, educated me, worked with me, gave me much time for interviews, and provided me with copious amounts of information and commentary. My success as an industry journalist is in large part due to your contributions.

Clare Goldsberry
Phoenix, AZ
2011

Chapter 1

Background – the evolution of the custom injection molding business

As the plastics industry has evolved, so has the business of injection molding. New and varied materials, new machinery technology, new mold technology, and literally thousands of new applications have increased the complexity of business for the custom injection molder. Additionally, with globalization and competitive issues assailing molders from all sides, the business has become one in which many varied skills, knowledge sets and talents are required to grow the business and be profitable.

Each one of these has impacted the business of injection molding and the ability of molders to be profitable in an age when recession, rising prices, tightening credit and shifts among OEM customers (original equipment manufacturers) bring uncertainty to an industry that at one time was fairly stable. Sixty years ago, when molders were constrained by machinery technology, fewer materials and even fewer applications, molding was an easy business in which make a living. One could buy an injection molding machine, teach oneself how to operate it, find something that fitted the bill for plastics, and be successful.

Case history: Landis Plastics

One of those success stories in the United States was Landis Plastics, a family company whose growth over five decades and four generations was nothing less than phenomenal. From a proprietary molder of plastic wall tiles as Landis Tile Co. in the 1950s, the company evolved into a custom injection molder of coffee can lids in the 1960s (the arrival of ceramic wall tiles from Europe was the 'disruptive technology' that killed plastic wall tiles).Then for the next three decades, making a plethora of packaging products for most of the Fortune 100 food and consumer goods companies, and using some of the best molding and decorating technologies available, Landis reached its pinnacle. The company was sold to packaging giant Berry Plastics in 2003 (**www.berryplastics.com**).

There are many stories similar to the Landis Plastics saga, not only for the many family-owned plastics businesses in North America but Europe as well. Those companies that evolved successfully from small, one-customer or one-product companies with two or three machines into multi-national injection molding firms, did so through understanding change, adapting to change, and adopting new technology.

Along with that evolution, the business of injection molding changed. Many years ago, Merrill Roth, a partner founder of custom injection molder Grant & Roth said it best: "*Today, molding is big business. Time was we didn't know anything about business. It was all just trial and error until we got a business going. We even built our own presses because we couldn't afford to buy them.*"

1.1 Change in materials technology

Much of today's injection molding was made possible by materials development that took place during the high-growth years of the 1970s through the 1990s. Polypropylene and polyethylene, considered 'cutting edge' for some products in the 1950s are now commodity materials. Driven by a desire to exploit plastics in a variety of new applications, engineering thermoplastic materials were developed for markets such as automotive, medical, packaging and toys. But perhaps the greatest advances in plastics came in the 1980s when the personal computer hit the consumer market in a big way, bringing almost unprecedented demand for engineering thermoplastic parts molded from ABS, polycarbonate (think the ubiquitous Lexan developed by GE Plastics), and other resin compounds.

New materials development allowed the automotive industry to reduce the weight of vehicles and save on fuel consumption, something that became a major focus in the early 1970s during an oil shortage. Today, some 70% of the components in vehicles, including passenger cars, light trucks, and even heavy duty tractor-trailer trucks, are made from plastics – from under-the-hood, parts molded from high-heat, chemically resistant engineering thermoplastics, and fuel systems components, to exterior bumpers and door trim, and interior consoles, instrument panels, door panels, steering column covers and much more – plastics contribute to strength, light weight and aesthetics.

Ongoing material developments throughout the 1970s and 1980s provided an avenue for all types of products in every industry and market segment to be converted to plastics, providing injection molders with big opportunities for business as the transition from glass, paperboard, aluminum and even steel to plastics gathered steam. Plastics were finally coming into their own, and the injection molding process seemed to be answer for literally thousands of products.

1.2 Change in molding machinery technology

More than any other single thing, computer technology in the 1980s changed the molding machinery business, and the business of injection molding. Molding actually became a technology, a true science, rather than an 'art' that depended upon the expertise of one person on the shop floor. While human expertise is and always will be necessary to the success and profitability of the injection molding business, computer aided technology brought a level of precision to the various parameters of molding that hadn't been available before.

Dealing with the multiple variables in molding (i.e. temperature, pressure, etc.) has always made the difference between optimum cycles that give a molder profits vs. cycles that result in breakeven or worse. Precision with computer-aided process controls gave molders an opportunity to improve the entire injection molding process and actually fine-tune profitability.

Electric molding machines also provided something new in the way of productivity and efficiency. While some molders still love hydraulics, the all-electric machines caught on with many molders looking for better repeatability and energy savings that the electrics offered. During the boom times of the 1990s, many molders went to all-electrics to maximize their productivity, and to reduce their energy consumption.

Automation came into the picture in the 1980s, with pick-and-place robots used to some extent, and by the 1990s robots were being added to most molding machines used for high-volume, high-speed molding applications such as packaging items, medical disposables, in fact almost anything that came out of a multi-cavity mold and needed to have aesthetics preserved, be put in a box for shipping or taken onto a conveyor for downstream operations.

Decorating operations such as hot-stamping, pad printing and painting provided additional revenue streams. Other secondary operations also began in earnest such as assembly, sonic welding, sonic insertion of metal bosses for screws and other manual steps that also added value and created more revenue. Molders began realizing that there is more to making money than making parts. Value-added activities can bring greater profitability, something that would carry forward into new technology in the new millennium.

As more and more products moved into plastics, more injection molding facilities came into being to meet the need. Custom molding plants are 'job shop' facilities that take in molding projects from all segments of industry. Many of these job shop molding businesses specialize in various markets or in types of products, or the size of the parts that need to be molded. Many molding plants came onto the scene with smaller molding machines, under 300 tons primarily. Some had equipment that went up to 500 tons, but few custom molders wanted to go over the 500 ton range in equipment because of the increased infrastructure that was required, such as overhead cranes to lift the molds in and out of the machines.

1.3 Change in products and markets

Custom molders comprise a good portion of the injection molding business, with the proprietary molding operations of some large OEMs in the packaging, medical, pharmaceutical and electrical sectors effectively making up the balance. While there seem to be no accurate statistics with respect to the custom vs. proprietary businesses, in terms of the weight of material used, it would appear to be split about equally. For custom molders, their customer base is varied, the types of products and components they mold are also varied, and so are the materials they use to mold these products.

Case history: Deseret Pharmaceutical Company

It was in the 1960s that medical devices, particularly syringes and other disposable items started to move towards plastics. Some of the early pioneers in this industry were James LaVoy Sorenson, Victor Cartwright and Dale H. Ballard. Sorenson was a pharmaceutical salesman in Salt Lake City, Utah, and sold pharmaceutical supplies to doctors. He saw first-hand the problems doctors were having with their medical devices such as syringes that depended upon repeated sterilizations to maintain cleanliness, as the same device was reused on multiple patients. Plastic materials were new at the time, but Sorenson became convinced that medical devices would be prime applications for plastics. He teamed up with Cartwright and Ballard, and the three soon became well known for their contributions to converting metal and glass medical devices to plastics. These three men, all from Salt Lake City, founded Deseret Pharmaceutical Company, a maker of disposable medical devices. To this day, many of the companies in the large and thriving medical device industry in Salt Lake City can trace their roots to Sorenson, Cartwright and Ballard's efforts to convert many medical devices that required sterilization into disposable plastic devices. Today, the medical device industry is one of the largest consumers of plastics, and while many medical OEMs do their molding in-house, a large number of custom molders supply that industry, with the medical sector representing the fourth largest end market according to several estimates.

During the 1980s, many of these custom injection molding companies realized their best opportunity for growth lay with the booming personal computer industry, personal video games (Atari was one of the first manufacturers), and the growth of office equipment such as copiers, printers, and other equipment that came with the computer industry. This market swept the plastics industry along and provided new opportunities for materials development, molding and mold technology. It was a time when nearly every custom molder one spoke with was doing some type of computer and/or business equipment part for IBM,

Hewlett Packard, new startups such as Compaq Computer, Dell, Lexmark (an IBM spinoff) or Apple, and many more, smaller companies that soon became much larger.

While those two markets represented the best opportunities for custom molders, other markets also followed along such as the packaging industry, consumer products, and home electrical and electronics (lighting and electrical products, and consumer electronics such as televisions, personal music players, mobile phones, etc.)

The automotive industry rapidly became a major consumer of plastics, although the US molders and moldmakers that served that market were mostly located in the 'Rust Belt' – that section of the Midwestern United States where much of the Industrial Revolution had been located. In addition if you were an automotive supplier then you were likely to be exclusively an automotive supplier because that industry was so demanding with its qualifications, certifications and many quality requirements that required suppliers to maintain a large infrastructure in terms of personnel.

By the early 1990s, there were so many custom injection molding companies that competitive pressures caused many captive molding operations to close, and the OEMs began transferring the work to custom molders to reduce costs. That was particularly true for the medical industry where nearly all medical companies had initially set up their own molding facilities because of the level of control required to maintain good manufacturing practices, including in many cases clean room molding facilities and FDA oversight. Other 'high-tech' industries also began divesting themselves of in-house molding, such as the electronics and computer industry. These large OEMs had typically reached a crossroads where they had to make decisions about capital equipment expenditure – of which molding equipment was a huge part of a company's budget – against investing in new product R&D, marketing, sales, advertising and other OEM activities that could generally only be done in-house.

So the medical and other high-tech companies began divesting themselves of in-house molding and contracting with custom molders. However, they typically wanted more than just molders of parts, demanding additional value-added services which provided expanded opportunities for molders. In response to this demand, custom molders became contract manufacturers, equipped to take on more of the downstream operations such as hot stamping, pad printing, sterilizing and packaging.

At this point in the custom injection molders' history, many had become technically very adept, developing broad capabilities in not only molding but a variety of secondary operations required by the large OEMs such as plating, painting, shielding, assembly and decorating. If the molding companies didn't have these capabilities in-house, they worked with subcontractors in these areas, particularly in painting, EMI/RFI shielding and plating, which tend to be specialty services.

The 1980s and 1990s saw the rise – and fall – of many exciting new business areas, primarily in the consumer electronics industry. Molders were confronted with the new reality that diversification is a good thing, and that being able to 'shape-shift' into a supplier that can meet a variety of demands for services would keep them growing. This was true in many areas, including the compact disc (CD) business and mobile phones, amongst others.

The late 1980s and early 90s saw the rise of the CD industry. Not only did that mean molding of the compact discs themselves (which was primarily done by the big record companies in proprietary facilities), but the packaging for all these millions of compact discs (mainly in 'jewel boxes') became big business for custom molders. The CD industry was the disrupter of the tape cassette business, which had also been a good business for custom molders, but declined rapidly. Several specialty custom molders built fast-growing businesses in CD packaging, as well as the video tape packaging business. That run lasted about 15 years before the advent of DVDs, MP3s, downloadable music, and 'movies on demand' from satellite and cable TV, which started a steady decline for the CD and video tape markets, much like the vinyl record business fell with the rise of the cassette tape and then the compact disc technology.

Many molders in both the US and Europe were growing business with the cellular telephone industry, as companies like Ericsson, Nokia, Motorola and others were making cell phones faster than people could purchase them. New models came out regularly, as consumers demanded more and more applications, and these mobile phones moved from the large 'brick' phones of the early 1990s to the small phones we enjoy today. However, by the late 1990s, the market was becoming saturated, and not every player in that field was realizing success.

One large custom molder that made thousands of cell phone housings for Motorola found itself sinking when that company's products began lagging in a consumer market where tastes were rapidly changing. At one point, this molder had three truckloads of molded cell phone housings sitting in the parking lot, the shipments having been rejected by Motorola. Some molders couldn't respond fast enough to these rapidly changing consumer electronics markets to stay in business, and by the turn of the new millennium, many had fallen out as the bulk of that business went first to lower cost countries such as Mexico and finally to China.

Still, as recently as the mid-1990s, there seemed to be a never-ending cycle of new products coming to market, and custom molders responded with more plants, larger plants and more machines. Machinery manufacturers couldn't make machines fast enough to keep up with demand from the molders, who were booking work so quickly that many couldn't wait the eight weeks it was taking to have a new machine shipped.

1.4 Change in captive and proprietary molding

Captive and proprietary injection molding, while still prevalent in some industries, aren't as big as they once were. As the custom molders grew in size and also in the scope of their capabilities, expertise, experience and processing technology, those companies that molded their own products began to relinquish that part of their manufacturing business to the companies whose business was molding – and who could do it better - the custom molders. OEMs began focusing on what they did best: product development, marketing, sales and those functions within their business that made them money rather than cost them money in terms of capital equipment.

Sometime in the mid-1990s, while touring a large medical OEM in Southern California, their molding supervisor apologized for the old equipment, noting that the OEM saw the molding operations as one of the least desirable parts of the company. Molding, while essential to making their products, was not core to the company's business and therefore when it came to getting new equipment, upgrading the molding operation, and hiring people with expertise in molding, the company wouldn't make the investment. He was left to run the molding operation the best he could with what he had. It wasn't long after that visit that the company began divesting itself of its molding to custom molders who had clean room molding environments and the capabilities to run the products.

Molders that were captive to one or two very large companies sensibly became less enamored with 'keeping all their eggs in one basket', and spread their wings to take on other work and diversify their business model; those that didn't sometimes found themselves in trouble with too much dependence upon one company.

Many proprietary molders found that to make a good return on investment with their molding equipment they had to enter the custom molding business. Maxell enjoys a worldwide reputation as a leading brand of electronic storage media and battery products. However, its injection molding facility in Telford, UK, produces moldings for food packaging, pharmaceutical, automotive and many other industries, as well as design services, assembly, printing and warehousing and distribution (**www.maxellmms.com**). Keeping the machines running is always the primary concern for molders, and it was no different for proprietary molders. However, it increased the competition for custom molding work.

1.5 Change in geographical locations

The injection molding business typically grew up as a 'local' business, with owners tending to be 'family' and with one facility located close to home. With the increasing opportunities in the computer, business/office equipment, and cell phone OEMs, molders began looking at other locations from which to serve these large corporations. As these large OEMs began setting up plants in various parts of the US for example, molders were asked to set up a facility to support that plant with plastic components that didn't need to be shipped across the country. Most molders responded positively to that request.

The computer and consumer electronics industries were clustered in cities like San Jose, Santa Clara, and other 'Silicon Valley' locations in California. Compac Computer was located just outside of Houston, and Dell Computer Corp. in Fort Worth, also in Texas. Many times it was a matter of new companies being spun out of larger companies, such as when IBM spun off its printer division to become Lexmark International, based in Lexington, Kentucky. In other cases a few management employees from one large OEM came up with a 'better mousetrap' and opened their own competing company in the same area, which caused certain regions to grow rapidly in the manufacturing arena and attracted molders in the process.

One booming region of the US during the 1990s was the Pacific Northwest (Washington, Oregon, and into Idaho) where companies such as Hewlett Packard, Epson Printers and others established large manufacturing plants. While there had always been a number of molders in that region – most of them small, family-owned businesses – the expansion of manufacturing provided huge opportunities for larger, multi-plant molding companies that were growing their operations to establish a plant in the Pacific Northwest and take on more business. Ultimately, that region became a hotbed for manufacturers and in turn, custom injection molding companies. Some of the larger molding companies that set up facilities in that region included The Tech Group (Tech Plastics, the custom molding division, is now owned by West Pharmaceutical, **www.techgroup.com**) and Nypro (**www.nypro.com**).

Today however, molding in the Pacific Northwest is a shadow of its former self. The heydays of the 1990s are gone, as are most of the large OEMs and many of the molding companies that took that region by storm. Left are a dozen or so small-to-mid-sized molders that manage to serve what's left of the manufacturing community in that area, with some measure of success.

1.6 Globalization

Both OEMs and their suppliers in the established manufacturing regions of the US and Europe have had to respond to a much greater challenge affecting industry and business worldwide – that of globalization.

The remarkable growth of Asian and particularly Chinese manufacturing, at the expense of more traditional producing countries has been well-documented. However, globalization began impacting the North American plastics industry at least two decades prior to the 1990s, with many large OEMs, particularly those in the automotive industry, putting plants in Mexico. Mexico was an attractive option: it was close to these companies' primary markets (the United States and Canada) yet they could take advantage of the lower manufacturing costs, especially manpower. Twin-plant 'maquiladora' operations enabled some of these OEMs to manufacture in Mexico whilst keeping administration, warehousing, and other management operations on the Texas side of the border, and molders who were reluctant to cross the border could initially serve these companies by locating close to the Texas-based parts of their operations. Nypro (**www.nypro.com**), Summit Plastic Molding (**www.summitplasticmolding.com**) and Tech Plastics Inc. (**www.techgroup.com**) all established molding facilities in El Paso, Texas.

Gradually, some OEMs moved further into Mexico to places like Monterrey and Guadalajara (which became Mexico's 'Silicon Valley'). When Guadalajara became a booming manufacturing region, many molders couldn't resist the siren call to follow their customers. The OEMs such as Eastman Kodak and Hewlett Packard wanted their suppliers next door, putting in large manufacturing campuses that could accommodate a dozen or more buildings to house their suppliers. Some suppliers followed such as Tech Group Inc. and others. But many of the OEMs wouldn't stop for long in Guadalajara. No sooner had some landed in Mexico, than they were already searching for the next low-cost country, and found in it Southeast Asia. Singapore became popular first, but by the mid-1990s as relations between China and the US became friendly, China opened its doors to foreign direct investment and the flood began.

A number of molding companies including The Tech Group and Nypro soon found themselves following their customers once again. Nypro hung on for some time in the Pacific Northwest, finally closing its Corvallis research and manufacturing in 2009, as well as selling its Chihuahua, Mexico facility the same year.

Nypro's moves around the globe became the model for many larger molding operations, yet it remains a privately held, employee-owned company. The company still has its headquarters facility in Clinton, Massachusetts, and plants in Chula Vista, California; and also Tijuana, Juarez, Guadalajara and Monterrey, all in Mexico. Today, Nypro also has locations in China, and other countries globally as a means to be close to their customers.

Much the same thing was happening in Europe as companies began looking beyond the mainstream cities of France, Germany, Italy, Austria, the UK and Netherlands. They initially established manufacturing plants in the lower labor-cost countries within the European Community at the time, drawn in partly by financial incentives. Ireland experienced particularly strong growth, with medical OEMs and business machine manufacturers such as Dell creating business for incoming molders. Subsequently, after the fall of communism and the enlargement of the European Community, much more molding work moved east to countries such as Poland, Hungary and Romania, which were keen to attract foreign investment and good jobs for their people. Ultimately, European companies too found themselves competing with the manufacturing power-base in Asia.

However, it wasn't just the molding business which moved around the globe. As long ago as the 1980s US molders – particularly those on the West Coast – were hearing talk of Taiwan as a low-cost source for molds, and many found they were quoting against Taiwanese moldmakers (and molders) for the booming computer, telecommunications and consumer electronics business. Elsewhere, and especially in Europe, the new low-cost country for obtaining molds was Portugal. Ultimately much moldmaking also moved to China and other Asian countries.

Ultimately, Asia and the Far East have had a huge impact on both North America and Europe, with places such as China, India, Singapore, Malaysia, and Thailand currently the low-cost countries of choice at the beginning of the 21st century, taking manufacturing plants and jobs. China has become the giant in manufacturing making increasingly high-tech products, in spite of some concerns regarding product quality and safety, and also taken much of the moldmaking. Machinery manufacturing still tends to take place wherever the machine will be sold and used, due to the cost and inconvenience of shipping heavy plant.

The development of the internet and e-commerce has forced companies to compete in a global market, but also provided manufacturing companies with ways to facilitate offshore manufacturing. Many companies in both the US and Europe hire mold design and engineering firms in places such as India, known for their low-cost engineering talent. To companies in the US for example, the designers work while the US is sleeping, so when the mold shops' doors open in the US, designs are awaiting approval.

Quality continues to be a huge issue with global manufacturing, particularly with respect to mold manufacturing in places such as China, where much of the moldmaking has gone. Lack of control over the type of steel used, the types of components such as hot runner and valve gate systems and other components required for a high-quality, high-production mold, mean that molds often arrive in the US or Europe already needing extensive rework and repair. However, some mold companies have actually made it most of their business to perform the rework and rebuilding of molds coming from China. They take lemons and make lemonade.

In addition to the quality issues, problems related to intellectual property have also plagued some companies that take their moldmaking offshore. There have been well-documented cases where a moldmaker will make one mold for the customer and another for themselves, and begin producing the products for export and sale.

Shipping costs and the global environmental carbon footprint are also providing cause for rethinking the offshoring of molds and molded products, particularly for molds that will be shipped back to molding plants in the US or Europe. With the huge amount of goods now being shipped from Asia, container ships can also be in short supply and as the price of fuel has risen, manufacturing in Asia has certainly become less attractive.

1.7 Change in the manufacturing and business environment

Over the past two decades, we've seen the boom and bust of injection molders and mold manufacturers in North America and Europe. The new millennium brought many changes in the way all companies have to conduct their business, albeit with the fallout of many of the early internet pioneers. As the whole world became visible competitors through the internet, China loomed large on the plant floor of many molding and mold manufacturing plants, and with their dominance, the United States business fell into a small recession. Molders and mold manufacturers in Europe and the US scrambled for ways to become more competitive with China, India, Singapore and other southeast Asian countries, but continued to see work leave for the cheap molds and molding of the 'low-cost countries.' The LCCs, as they became known, took precedence over the more expensive countries that had long been known for their innovation, creativity, quality and productivity in things such as mold manufacturing and molded components.

> ☞ **The new millennium brought many changes in the way all companies have to conduct their business.**

However, new competition isn't all bad. It drove molders to find new ways to increase efficiencies and productivity, to become more savvy business people and find ways to sharpen their skills in marketing, sales and quoting. The industry wasn't going down without a fight, and while many molding and moldmaking companies have been driven out of business there remain many stronger and better companies. Today, many of these companies look very different from their 1990s selves, when they couldn't mold parts – or buy equipment – fast enough. No longer accidental molders, they are now molders by design with strategic plans and purpose.

No one knows the precise number of injection molding companies in business throughout Europe and North America. While we know there was a great deal of fallout in the industry, there was also much consolidation primarily among the proprietary molding companies and particularly in the packaging market segment. However there are many more proprietary molding operations that still exist behind the walls of large OEM manufacturing operations.

According to the latest report from Applied Market Information Ltd., Europe's injection molding business lost almost 23% of its value in 2009, much of that due to the global economic downturn, falling to 42.6 billion Euros ($56.4 billion US) in 2009 from 55 billion euros ($78.8 billion US) in 2008. The decline was not evenly distributed, with the UK suffering most. Polymer demand grew in Central and Eastern Europe as capacity continued to shift from west to east, and more than 12% of injection molding sites in Western Europe closed, with further closures still anticipated. Relocation of capacity outside Europe also continued. However AMI predicted that the industry overall would return to growth in 2010, forecasting an expansion in polymer demand of 3-4%, driven by development in packaging applications and specialized sectors such as medical molding.[1]

Statistics compiled for the American Chemistry Council show a major fall in thermoplastic resin sales to all major market sectors between 2005 and 2009.[2] Total 2009 sales were almost 15% below the 2006 peak, although certain lower-volume materials were excluded from the 2009 totals. All individual markets saw a fall (the biggest – packaging – held up better than most) except for the only 'sector' which saw an increase over the period which was 'exports', demonstrating the trend towards globalization and offshoring described above.

1 The future of the European injection moulding industry, Applied Market Information Ltd., Bristol, UK, **2010**, www.amiplastics.com.

2 Thermoplastic Resin Sales by Major Market, 2005-2009, American Chemistry Council Plastics Industry Producers' Statistics Group, **2010,** as compiled by Veris Consulting Inc., www.americanchemistry.com.

However, at the time of writing, there appear to be some signs of an upswing in manufacturing, as inventories have fallen off and consumer demand has picked up. Both North America and Europe seem to be recovering somewhat in their manufacturing sectors. While no one has a crystal ball, plastics is too much a part of nearly every market segment to be down for long, as we're starting to see.

Managing your business

Chapter 2

Developing marketing strategies for profitability

2.1 What is marketing?

> ☞ **Marketing is about creating demand for your products and services amongst those customers and companies that value them and are willing to pay for them.**

Marketing remains an ambiguous word for many people in the custom molding business. It's typically thought of as something that the big OEMs engage in, such as marketing a new model car or a new brand of toothpaste to a consuming public. However, molders have a consuming public – it's just a different type of public. So how do you market to that public when you're not really marketing a product?

Marketing is about creating demand for your products and services amongst those customers and companies that value them and are willing to pay for them. It's about promoting your strengths. It's about selling **value**. That is the way you become profitable, and the customer gains a quality, reliable supplier. Since most molders do not have products – with the exception of those who perhaps have developed proprietary products, they primarily sell services. These will increasingly include moldmaking, building manufacturing cells with automation, secondary operations, new product development assistance, assembly, packaging, and other services that come with being a contract manufacturer in today's molding world. Many of these are '**value added**' services – those services that go beyond just molding parts and putting them in a box, and can generally increase the price that a molder can charge his customer.

While molders can certainly offer a wide range of services to a variety of customers in diversified industry segments, it isn't possible or even desirable today to be all things to all people. You must discover what it is you do best, who you do that best for, create the demand for those things among customers who value what you do, and then do it better than anyone. You must uncover your '**core competency**' and market that to those OEMs who can best benefit from that core competency, and create a demand for your services among that group. Why? Because it is the companies in that group who will pay for what you have to offer!

2.2 Creating a marketing strategy

One of the keys to preventing companies from remaining 'accidental molders' is creating a marketing strategy. Marketing is key to profitability, and while it differs from sales, marketing is a key component to creating a sales strategy, as we will see. Marketing is comprised of activities that create demand for your products and services. And while this involves research, the ultimate goal is to develop a marketing strategy that will create demand for your services and develop new business. While many molders in business today began their companies several decades ago, capturing work from a variety of companies and developing niches in various market segments, remaining in the business of injection molding in the 21st century can no longer depend upon luck. Flying by the seat of one's pants worked in the 1960s and '70s, and even into the 1980s and '90s, but that business 'strategy' has run its course.

A number of years ago, John Weeks, (then President of Precise Technology Inc., now Rexam), said the following: "*. . . I suggest that some of our best molders, and indeed most molders, who excel in technology and disciplines, do diminish or dismiss the importance of marketing and sales – for the very reason they excel in other areas. They believe they make a better mousetrap and therefore the world will come to them.*"

It's doubtful that many believe that today, after the first ten years of the millennium – and especially the recent global economic recession – have left many bruised and battered, and struggling for more business and new business. Still, while most understand the nuts and bolts of their business – machinery, molds, processing and the like – there are still a number of molders who do not understand the less tangible aspects of business such as the development of a business plan, marketing and sales strategies, or even the finer points of a profit and loss statement.

Today, to be successful and profitable requires a strategic marketing and sales plan; one based on sound and ongoing research into three main areas: 1) knowing your own company, its technical capabilities and human expertise; 2) knowing your customers and their technical capabilities and human expertise; and 3) knowing your competitors. This last area is one that many people leave out when thinking about a marketing strategy. Yet, if you don't know your competitors, how can you compete?

This information gathering (research) exercise is only the first stage of building a marketing strategy. The second stage is to analyse the information you have gathered, and the basic principles of a process called a SWOT analysis are described later in this Chapter. **SWOT stands for strengths, weaknesses, opportunities and threats**. Finally, if your efforts are not to be wasted, it is essential that you act upon the outcomes of your analysis. The final chapter of this book is a case study based on a survey of a real company, which shows the conclusions reached, and subsequent recommendations.

2.3 Know your company

> ☞ **You can't possibly know what markets and what companies within those markets to go after if you don't know what your company does best and who it does it best for!**

Your goal with this exercise is to closely evaluate your company, along with your current customer base and your competitors, and to identify those things that differentiate you from your competitors, and those capabilities, areas of expertise and technology that makes you successful with your current satisfied customers. Being another 'me too' molder doesn't cut it in today's fast-paced competitive environment. No company can be all things to all customers, and to try that strategy in today's economy will result in being not much to the wrong customers.

You can't possibly know what markets and what companies within those markets to go after if you don't know what your company does best and who it does it best for! In fact, not knowing the difference between what you do well and what you don't do well can often lead to trouble. Companies who do not plan their strategy beginning with this evaluation often end up taking work that is beyond their area of expertise. A down economy is the main time when molders are tempted to take on work they probably shouldn't. A desire to break into a new market is another reason many molders take on jobs that are outside the scope of their capabilities and expertise. **You really can't afford to practice on a new customer with a new project that pushes the limits of your technology, your capabilities and your expertise.**

Many years ago, the company for which I was selling molds and injection molding was low on work. I'd got an opportunity to quote a job that involved a number of large, multi-cavity, unscrewing molds that were actually outside the range of our molding capability at the time. However, when business is down and a big opportunity comes knocking, it's often difficult to turn it away and the desire for business prevails over common sense.

It didn't seem to matter that the company, while having built many complex multi-cavity molds, had never built an unscrewing mold. That should have been our first warning. We were about to venture beyond our expertise. The project ended in disaster, the company lost thousands of dollars on the project and we ended up relinquishing the job back to the OEM unfinished so they could find a vendor that was adept at unscrewing molds. The lesson I learned from that experience was clear: don't practice on new customers!

Every company has a forte or a 'core competency' – or at least it should have! If you don't believe you have a core competency, do this exercise and generally you will discover that you actually do, and it will come to light as you proceed. Maybe you do really well making multi-cavity molds, and molding small, intricate parts that weigh only a few grams. As you review your jobs records, you might find that every time you build one of those molds, you are successful and profitable, the mold is on-time and on-budget and the customer is thrilled with the outcome.

Perhaps you do really well at large single-cavity molds for automotive instrument panels or other large components for automotive applications. That might translate well into large components for agriculture products such as tractors, lawn-mowing tractors, snowmobiles, all-terrain vehicles and other types of vehicles that have parts similar to the automotive OEMs.

When molders actually sit down to strategic planning, they're often surprised to find they really do have a core competency, a niche in which they've been successful, but have just failed to recognize that fact. It's all in the way molders think of themselves, and some still think of their business as a 'custom' molding business that involves doing a little of this and a little of that for a wide range of customers and markets. Many still have an 'opportunistic' mentality – "*If I take on any work that comes my way, some of it is bound to be good (profitable) work.*" They try to play the odds, but let's face it, this is business not poker!

Money made on a profitable job never makes up for money lost on another. Low-balling a mold or molded parts is just plain dumb. There's an old molder's joke that says, "*I'll low-ball the piece part price and make it up in volume.*" That type of flawed thinking will get you in trouble every time. That's why it's critical that you perform this evaluation on your company. Find the winners – then become determined to find more of those. Look at the losers, and strategically begin planning to help them find a new molder, and find new work to replace the losers with more of the winners.

As a whole, the custom molding industry tends to be extremely fragmented, as thousands of companies mold parts for thousands more companies in dozens of markets. This fragmentation is reflected at the individual molding company level. Lack of focus is a major downfall among custom molders, and doing this evaluation exercise will bring your focus back to your company; and help you to home in on those things that result in profits and successful business relationships.

If you took a look at a typical plastics industry trade directory, you would notice one thing about those companies listed under the injection molding section: The largest molding companies are those focused on one or two markets. The first 10 companies typically only have one focus – generally automotive or packaging. Now, keep going down the list and notice that the smaller the molding company the more markets they list as those they serve. I've often wondered how a molder with sales of $1million a year can possibly be serving every market listed in the key! Yet some of them list upwards of 20 markets. That is impossible to do successfully. These opportunistic molders will always be small companies, grabbing at any job that comes along, and will get it based mostly on price.

Focus means that you know what you do and how to do it better than anyone else. "*Focus your sales and become the kingpin in your niche so that everyone wants you,*" said one industry consultant a number of years ago.

That's why the first step in formulating a good marketing strategy is a thorough examination of your business. You must know where you are now before deciding where you want – or need – to be in the future. To guide you through this process, you'll find template questionnaires at the end of this Chapter. These can help you begin to identify your strengths, weaknesses, opportunities and threats – what is called a SWOT evaluation. This will help you determine those things that you do really well from those that you don't do well, and also the types of customers that fall into each category. It's not uncommon to hear a molder say, "*We always fall flat on our face with that customer.*" If you find yourself saying that, you need to find out why and perhaps become more strategic in your thinking.

2.3.1 Know yourself – surveying employees

> ☞ **Every company has its soft spots.**

An excellent way to get to know your company, its strengths, weaknesses, opportunities and threats is to survey employees. Using the template questionnaire at the end of this Chapter as a guide (Appendix 2.1), it is good to ask a cross-section of employees, not just upper management. Employees such as customer service personnel, the receptionist, the warehouse supervisor, the lead operators, etc. will frequently see any problems at first hand. They will all provide valuable – and different – input that you will find extremely helpful. The questions are also designed to get employees to think about the company as a whole, not just to air their complaints about someone in management or a co-worker.

As important as it is to define your strengths, it is equally important to determine your weaknesses. Ego can sometimes get in the way of company management admitting to any weaknesses, but every company has its soft spots. Identifying these weaknesses allows you to see where you're vulnerable to losing money – or customers. Are there poor communications within the company that permit things to fall through the cracks? If so, the customer is usually the first to notice.

Knowing your weaknesses also helps you to evaluate those things you're good at vs. those you're not. Perhaps your company is always on schedule with large, long-running production jobs, but always late with the few thousand parts a customer wants molded every three months. This could signal that your forte is molding high-volume, long-running jobs. At this point, you need to decide whether to jettison those jobs in which you are not as successful or profitable, and stick to those jobs which you currently do well. Alternatively, is there a creative way to do this type of work that lies outside your forte, such as dedicating a machine or two just to short-run jobs?

Employees are an excellent source of knowledge with respect to what's actually going on in the various departments. While the owner or the company president may think they know exactly what's going on and why, I know from having done many of these surveys within companies that you probably do not. In fact some employees will tell you more than you want to know.

Another helpful idea with respect to surveying employees is to engage an outside third party to ask the questions and interview those chosen for the survey. Employees will be more open and above board with someone who is not in management at the company. Also, if the person giving the answers is kept anonymous they are generally more comfortable in giving honest, open answers and opinions. Besides, all upper management needs are the results of the survey – they don't need to know exactly who said what. You can hire an outside marketing or PR firm with the capability to conduct the survey. Alternatively you might find a marketing student from a local college who could take it on as a project or work experience.

While you may disagree with some of the responses of those surveyed, it is important to remember that there are no right or wrong answers, and all opinions are valid. Perception is reality, after all, and if an employee has a perception of the company, that perception is their reality. And if it is one person's perception, it is probably the perception of others as well. Finally, giving employees the opportunity to provide input as to what the company does best, where improvements need to be made, and so forth makes them feel as if they have some influence on the success of the company.

Remember, the main point of the exercise is strategic – you want to understand the core values of your business. However, staff and customer surveys are also quite likely to flag up quality issues and weaknesses in basic customer service, even in the best-run companies (speed of response/delivery, failure to answer the telephone or emails, unhelpful staff, inconvenient hours or location, poor packaging, late invoicing) and this is all useful information – provided that these issues are addressed.

2.3.2 Know yourself – customer surveys

> ☞ **Not every customer is an asset.**

Surveying your customers is something that should be done periodically – once a year is what I recommend. Why? Because a lot can happen in a year! You need to get customer input so that you can benchmark your performance with current satisfied customers; to understand why they are satisfied – or to know what has happened to make them unhappy with your performance lately. From the customer's viewpoint, what are those things that your company is doing well? Where do you need to improve? Is the customer truly happy with your performance, not just in supplying quality molded parts but also in customer service, such as getting their questions answered promptly, and other aspects of your business. A template is provided in Appendix 2.2.

Your customer base is critical to the success of your company. It is also a good indicator of a company's true value. A customer survey is the next stage of knowing yourself, because you will learn a lot about your customers' business and what they value about you, and also gain some insight as to what to expect from them over the next year.

The survey will also help you decide whether or not a particular customer is the right customer for you. Over time, what may have been an ideal customer 20 years ago when they brought you that first, really great job, may have evolved into a customer that is no longer suited to what your company has become. For example, if you have grown into a molding company that specializes in multi-cavity, high volume molding, yet you're still doing some single-cavity, small volume work – say 1,000 parts a quarter – for a customer just because you've been doing that for 20 years, you may need to re-evaluate that business.

Not every customer is an asset, and to assume that the more customers you have the more valuable your company is can be a mistake. A molder with 100 customers may think he has tremendous value in so many customers when, in fact, less than a dozen of those represent any real source of growth and revenue potential, and any real value to his company. As well as asking yourself "*Who* are my customers?" you should be asking "Who *should* my customers be?" That is the purpose of knowing both yourself and your customers.

The value of a good marketing strategy is not only knowing who your customers *should* be, but who they *should not* be as well. Managing your customer base takes time and a lot of knowledge about each customer, and their actual and strategic value to your company. The actual value is the business they are currently providing your company, the revenues and profits that you realize from doing that work. Strategic value is that long-term growth potential for new business, or for more profitable, value-add business that can contribute to your company's growth.

The goal is to retain those customers who provide the best in actual and strategic value to your company. Those who drain your resources or that offer little or no growth potential should be let go by helping them find a new molded parts supplier.

2.3.3 Using the data you collect

This is where you start to compile your SWOT analysis, based on feedback from your staff and customers. You will have accumulated a great deal of data and opinion, and one of the hardest tasks can be deciding which category some of these things fall into. For instance, poor machine utilization could be perceived as a weakness, which is costing you money, and possibly making your prices to the customer too high. On the other hand, spare machine capacity might also be considered as an opportunity: you have the space to increase throughput without investing in another machine. The most important thing is that the information is recorded, analyzed and acted upon.

Under strengths, you should list all those things folks say you do well or most profitably. You may also list those things you offer that another molder down the street doesn't have. You may think that regular work from a market leading OEM is a strength in your business, or your location close to a major manufacturing center.

Under weaknesses you need to acknowledge where you could do better, or even which business areas you should be avoiding entirely. You might include ancillary and added-value services which you don't currently offer, or a lack of trained personnel. Negative customer feedback, data on failed quotes and information on high scrap levels and reject rates could all fall into this section.

Opportunities can be harder to pin down. It can be difficult to look outside you day-to-day business and see how and where it might grow. But maybe you have just done a small job very well for a new OEM customer – there could certainly be an opportunity there for building bigger and better jobs. Maybe you currently only work for one division within a major conglomerate – Could your good reputation open doors elsewhere in that company? Most importantly, look at all those factors you described as strengths – these all represent opportunities. If you perform really well for one customer or sector then you can do the same for another similar OEM or in a new sector requiring similar qualities.

Threats can also be difficult to identify, but forewarned is forearmed! Threats are also frequently beyond your control, and perhaps related to changes in markets and conditions. However, weaknesses can translate into threats, and intelligence from customers (or about them) can be particularly valuable. Again, the important thing is that you are aware of the business you are working in, and how it may develop.

A SWOT analysis is a great management tool for working out where you stand and where you should be going, and defining a marketing strategy and business plan for the future. Take a look at the case study in Chapter 17. There may be many parallels within your own business, and it will show you how to build on strengths and opportunities, whilst learning from the weaknesses and threats.

2.4 Know your customers

> ☞ **Knowing your customers' plans will help to ward off unexpected surprises.**

Provided you ask the right questions, your employee and customer surveys will also provide the information you require about your current customers. Each of your customers should be carefully analyzed by asking: "Who are my customers and why?" When you understand why your current customers do business with you – or why they no longer do business with you – you can more accurately nail down those potential customers you should be seeking out to grow your business.

Another reason to know your customers is to keep track of the market forces that are impacting their business that might result in a business downturn for you as their supplier, or an increase in their business that might mean more work for you. What acquisitions have your OEM customers made? What divisions have they divested? Why did they divest themselves of that business unit? Knowing your customers' business is critical to the success of your business.

Once I read an article in a business publication in which a large OEM had just been acquired by an even larger company. Knowing that a mid-sized molder in the Midwest did a lot of work for the acquired OEM, I called the president of the molding firm and asked him what impact the acquisition of his customer might have on his business. He was shocked to learn that his customer had been acquired.

Whatever impacts your customers is of prime importance to your business, and this is the main reason why knowing your customers is second only in importance to knowing yourself. The first decade of this new millennium saw heavy downturns in the telecommunications industry, with a glut of cell phones on the market. It also saw mergers and acquisitions in many markets including packaging and the appliance industry. Customers' requirements will change according to market share won or lost, or rise and fall with consumer demand, as we saw with the automotive industry in 2009. OEMs may decide to consolidate plants, reduce the number of suppliers, or move their manufacturing to another country to reduce costs. [See Chapter 7 – Markets and Profitability.]

Market forces that impact OEMs will ultimately impact molders and moldmakers, for better or for worse. If you are not aware of what is going on in the markets you serve, and with your particular customer base, you could be surprised, and we don't want surprises in business. When we are surprised, we react. In business it is always better to be proactive – not reactive to sudden changes in the wind. Knowing your customers' plans will help to ward off unexpected surprises, and the potential disasters these can cause.

By surveying your customers you can understand where you stand in their supply chain, and why. What value do you provide to your customers? Why do your customers do business with you vs. another molder, or why is your company among an OEM's stable of custom molders?

2.4.1 Using the data you collect

Based on the answers to your staff and customer surveys, **create a customer profile to help you determine exactly where each customer ranks within your organization, what percentage of your business each represents and what percentage of your sales, revenues and profits each represents**. If you have one customer that is 40% of your business, but represents only a small fraction of your profits, then perhaps you need to evaluate what you are doing for that customer that needs to change. Why are some jobs unprofitable? Why isn't that customer generating more revenue? Then separate the winners from the losers. Can you turn some of the losers around or do you need to let them go? How can you get more business from the winners?

If your customers work in a variety of business sectors you might also want to categorize your customers by industry to help you determine how much of your business is coming from which market segments – and are those markets seeing any vulnerability from outside forces?

For most molders, the 80:20 rule applies, which says that 80 percent of your business comes from 20 percent of your customer base, and that's a fairly typical figure for a custom molder. You may be surprised to learn that a small percentage of your customers support your entire operation. You may even feel uncomfortable about being so dependent on this small band. But what this also means is that just a few new 'ideal customers' could give your business a major boost. So what does the ideal customer look like? If you have analysed you customer base then you have the answer to that question, and that's a big advantage. It's different for every molding company, depending on those things that you've identified as your strengths – those things you do best – and for whom you do them best.

This information should also feed into your SWOT analysis. Look at your dependency on key customers. You may have difficulty deciding whether this is a weakness or a threat – perhaps depending on the growth prospects for a particular market sector. The most important thing is that once you have the facts in front of you, management have the data to support decisions and planning for the future.

2.4.2 Identifying your value to your customers

Once you've identified your strengths, you can begin to define these strengths in terms of their value to your customers, starting with current, satisfied customers because that is your core business. This is an important distinction because **people will not pay for what they do not value!**

For example, a molder once told me that he was thinking about putting in a clean room for his medical customers, who at the time represented about 30% of his business. I asked him if his medical customers were willing to pay the extra costs that would be added for clean room molding. He said he didn't know. I suggested that he survey his customer and ask them if having their parts molded in a clean room environment would be of value to them and if so, would they be willing to pay more for their parts molded in a clean room. So, we did a quick survey of his medical customers. Much to his surprise, he discovered that most of the parts that he was molding for these medical OEMs were components that didn't really require a clean room. So, naturally they were not willing to pay more for their parts.

Customers *will* pay for what they value. The CEO of one of the largest molding conglomerates in the United States was very adamant about this. Having a customer base willing to pay for the capabilities you offer is the sure path to becoming a successful molder, he said without hesitation.

While this might sound simplistic, have you ever lost a quote because of price? If so, have you queried the potential customer to see why they thought you were high in price? Maybe you have too many 'bells and whistles' in your plant? You wanted to build a 'Cadillac' mold when they only needed a family saloon? Understanding the problem is part of the solution – but maybe 'Cadillac' molds are your forte. Okay, so find those customers who *need* a Cadillac mold because they are the ones who will pay for that. When you find those customers who value what you have to offer, you will be able to move beyond price.

Fortunately, the molder who was going to put in a clean room found out what his customers valued before he spent the money to install it. This holds true for any service a molder offers. Your customers will not pay for any service they do not value. A good marketing strategy that includes a customer survey helps you avoid such pitfalls. Creating a profile of your best customers will also help find those business practices, policies and requirements in a potential customer that mesh well with yours, and good relationships can be developed.

The bottom line is that you want customers who are willing to pay for the things which you do well, and you need to offer the services which they value. Good marketing evolves from knowing your company and knowing your customers, and what they value in you as a supplier. Now that you have profiles of your best customers, you have a template to use when seeking out new business from new customers in new markets.

2.4.3 Marketing based on value

Now that you've laid the groundwork for your marketing plan, you need to incorporate the value model as a standard for acquiring new business. In the business world, a company either grows or fades away, and maintaining the status quo is not an option. However, defining growth and determining *how* you want to grow is key. This is where using the 'value model' comes into play – identifying value is key to planning a strategy for growth. While opportunity is important in any aspect of life, the custom molding business for too long was an 'opportunistic' industry. An opportunity to mold parts came along and the molder took it, with little thought as to how this job fitted into the molder's business. A business plan wasn't really a part of the picture, so there was no incentive to decline a job, or determine how it might benefit the molder in the long term.

Opportunistic growth however, is without direction and fraught with peaks and valleys brought on by the cyclical nature of your customers' businesses. Today more than ever, companies large and small are strategically planning their business, including looking at their supply chain and how that will be managed. **The choice of the custom molder is to manage or *be managed*!** If you do not have a business plan of your own, and a marketing and sales strategy to fit your business model, then you will very often find yourself at the mercy of your customers' strategies.

> ☞ **In the business world, a company either grows or fades away. Maintaining the status quo is not an option.**

The value model involves two primary types of value: actual (current) value, and strategic value. Actual value describes the value a customer is worth to you currently on a particular program, and the value you provide to your customers on current programs. Strategic value describes the value a customer represents to you in terms of potential, or future, business, as well as the value that you represent to your customers in terms of their strategic growth targets for the future. In both cases, the value model works both ways.

Your goal is to locate those customers with the highest actual and strategic value, and to be a supplier of value to those customers seeking what you have to offer in terms of actual and strategic value. Because you know that customers will pay for what they value, you must market yourself through the value model. You want to provide the services that are most important (valuable) to the customer.

If you sell value you won't have to play the price game. Ted Turner, owner of cable television channels and a very wealthy man, once said: "*The game is about building assets. I don't mind overpaying if I know the value is there.*" That is where you want to be – in a valued position with your customers so that you're not constantly haggling over price. In fact, if you *are* constantly haggling over price, you have not done a good job of selling your value.

But first you must work out what your customers value. Over the years that I've performed these surveys for molders and moldmakers, I've learned a few things about what OEMs value in their plastics parts suppliers. When purchasing agents were asked to rank the value of a mold on a scale of 1 to 10, with 10 being the most valuable, the average ranking came in at 3. In other words these purchasing agents didn't especially value molds, which meant they were 'price' buyers, not 'value' buyers. "*How much do you expect me to pay for a hunk of steel,*" one interviewee commented.

When engineers were asked the same question, they tended to rank the value of a mold somewhat higher – an average of 8. This suggests to me that because of an engineer's technical expertise, they understand that you can't get a perfect product from an imperfect mold – and that you generally get what you pay for. What this also tells you is that 'value' is in the eye of the beholder, and that people understand value in different ways. What provides value to a purchasing agent whose job it is to get the lowest price (and a big bonus at the end of the year for saving the company money), isn't what provides value to the engineer, who might have to deal with an underperforming mold, repairs, and maintenance costs – not to mention bad parts.

The RFQ bid-the-job game is often one of price. And it is a game! The OEM purchasing agents and engineers that most suppliers must deal with every day have been taught to play this game very well. They know that most of the molders and moldmakers they deal with tend to be smaller businesses that lack strategic plans, and many take advantage of that. But **providing quotations costs you time and money.** If a customer is using you as a quoting machine to keep other suppliers honest, you can choose to say no, and that you only bid on 'live' jobs.

An alternative solution introduces an element of value. If a potential customer with whom you have no track record inundates you with 'test' quotes, you can't charge for quoting, but you can charge for an 'Engineering Evaluation' on the component and provide them with your ideas for a mold design. By moving away from a game based purely on price, the client gets a better solution, you cover the cost of providing it, and hopefully you get the molding job too. But then stick to your guns – don't give that information away so they can take it to the molder or moldmaker down the street along with your ideas. It's time to let the large OEMs know that you are in business to make money and that you can't play 'bidding' games with them.

Your goal should always be to educate your customers about the value you provide, then provide that better than anyone. Additionally, you should maximize your value proposition to your customers by knowing what it is that your customers value about your company, and what they would like to see in the way of added value capabilities. Equally, you should minimize those activities that have little or no value to your business, such as spending hours and hours poring over dozens of RFQs, trying to quote every one of them, with little to show for your time and effort. The result of developing a marketing strategy will be that you will know which RFQs fit your business model and which do not; and also which RFQs are from purchasing agents 'fishing' and which are 'live' jobs that are worth your effort.

As mentioned earlier, the custom molding industry grew in opportunistic ways, but when you begin to think in terms of value, such as, "What is this customer's value to me and what value can I provide to this customer," you begin to turn opportunity into strategy. Opportunities strategically positioned in terms of their value become profits and growth for your company. "*Growth means capitalizing on opportunities*," said a marketing and sales manager for a major resin supplier in a talk I heard some years ago. And I might add to that: growth means capitalizing **profitably** on opportunities. What this means is that you don't completely ignore the opportunities that come along, but you quit being opportunistic in your thinking and become more strategic when examining them.

> ☞ **Building business relationships must be mutually beneficial. It is always a two-way street.**

Capitalizing on opportunities means that as you explore opportunities for growth, you look for those that maximize your capabilities, minimize your costs-to-manufacture, optimize your resources, and realize the most value to your business operations and to the business of your customers. Building business relationships must be mutually beneficial. It is always a two-way street.

2.4.4 New customers

If you follow the advice above, you should understand a great deal about your current customers and what they value about you, which will certainly help you to market your company to these companies and others like them. You will also know whether you are in the 'right' markets for your core competencies, and have a feel for whether these markets offer the growth potential you want for the future. However, we have already talked in the previous Chapter about the evolution of the molding business and the industries it serves. Nothing remains constant in business, and it is essential that you look beyond your current horizons from time to time. Customers come and go, and markets change. You need new customers who will also appreciate your value, and you may need to look beyond your current geographical markets or business areas. Chapter 7 provides more advice on identifying new markets, and a snapshot of the key opportunities for plastic molders. Chapter 4 also offers some useful guidance on sales strategies and targeting new customers.

2.5 Know your competition

Who are your competitors? One key to understanding how to place yourself in the custom injection molding market – and in the markets of your customers – is to know your competition. **With hundreds of competitors out there, even globally, you have to know your competition in order to know how to differentiate yourself from everyone else.** Many molders think they know who their competitors are, but few actually do. You might sometimes know who you are quoting against, and sometimes if you ask an OEM they may even tell you who else received an RFQ. However you really need more information than that if you are developing a strategic marketing plan.

As before, to determine who your competitors really are, you can start by asking your key employees in the 'Know Yourself' questionnaire. Then ask your customers. Many will tell you who their molded parts suppliers are, and will even provide other information such as why they value that supplier. However, unless you work in a very specialised niche, this is unlikely to give you the complete picture, especially in today's global marketplace.

AMI compile guides to the injection molding industry across Europe and also in China. These are regularly updated and based on direct contact with the companies they list. Company information includes locations, numbers of machines, polymers processed, quality standards and markets served (**www.amiplastics.com**). Current editions list more than 1000 sites across Europe, in spite of the recent market downturn. You can also get company information about molders from many other local directories and buyers' guides, as well as more general resources such as **www.thomasnet.com** and **www.kompass.com**.

Read industry trade magazines that often feature plant tours, company profiles and news about what molders are doing – who's buying who? Who is selling? Who has gone out of business? Who's buying new equipment for a big new program or investing in downstream added-value services? You might also get information from suppliers. Look at their websites, there is plenty of information at hand that can be readily found.

Attending trade shows, walking the aisles and looking at what your competitors are featuring in their booths is another good way to know your competition. Many molders say they have no time to get out of the plant. **I call this HIPS. No, not high-impact polystyrene. It's Head-In-the-Press Syndrome. Focusing entirely on your own company often means losing sight of what your competitors are doing, and what advantages they are offering.**

We're not talking about corporate espionage, but just knowing who your true competitors are and what they are doing is really key to differentiating yourself from them, or matching them in capabilities and services.

Knowing your competition is important in planning your strategy, as you sort out who you should be doing business with and who you shouldn't. If you're going after the business of a company, and you quote again and again to no avail (i.e. you're always too high in price) then perhaps knowing who you're quoting against will tell you that you shouldn't really be quoting this work at all. For example, Molder A, a mid-sized molder with 35 machines, couldn't understand why a potential customer wouldn't give him work despite all the quotes. Tracking the quotes, the molder found out that he was always too expensive. So, he asked about the successful bidder and found out that it was a molder across town who only had five machines and specialized in small, short-run production. Molder A learned he could never compete with a company like Molder B who had low overheads and a niche service, so he quit wasting time quoting the work.

Your goal is to quote against companies who are on par with your in terms of capabilities, services, expertise and technology. A mold company owner once told me that the best compliment he receives is to find out that another company – known as one of the best in the business – is quoting against him. "*I don't mind losing to a really good competitor, but it burns me up to lose to a small, low-tech shop that isn't close to being my competition, who low bids the job,*" this company owner said.

A scenario like that also says that many OEMs don't do their homework either, and often just scatter quotes out to every molder or moldmaker without even knowing their capabilities, services, size, technologies and manpower, or whether these shops are even capable of doing the job needed. But you can ask the OEM's purchasing agent or engineer which other companies are quoting, and increasingly importantly, are they looking overseas? That will tell you a whole lot about what you're up against, and if you're quoting apples to apples.

You'll always have competition and that's a good thing. Competition makes us better at what we do by challenging us and making business exciting. But your goal should be to make yourself impervious to the competition. **Remember, business is not about what you sell, it's about what the customer needs**. Too many times websites, brochures and other promotional materials talk only about what the molder or moldmaker has to sell because they have no idea what their target market really needs in a mold or molded parts supplier. If you sell only molding and your target customer needs a lot of value-added services such assembly, then you are not supplying what your customer needs and you open the door to molders who *do* provide that. If you are selling only moldmaking and have no way to sample the mold and do first article samples and mold qualification, then you are not supplying what the customer needs and are giving the competitors who have mold qualification capabilities a competitive edge.

If you believe that you are just a molder or a moldmaker, then your competition is legion. However, if you become a contract manufacturer and offer total solutions to customer problems, then your competitors become fewer. If you find a niche with special materials, or in specific markets then you narrow your competitors yet further. If you offer a range of service from new-product design assistance, prototyping and molds, through to molded parts, assembly and manufacturing cells, then you've blocked out a great deal of the competition who do not have that level of service and capabilities. You might also explore strategic partnerships with suppliers of complimentary services, such as a company that provides painting, plating or vacuum metalizing services; or one that performs hot stamping, sonic welding or other assembly services. This way you avoid the high entry costs of investing in new technology, while still offering a complete package.

> ☞ **Competition makes us better at what we do by challenging us and making business exciting.**

2.6 Marketing methods

Raising awareness of your company and creating demand for your services can be done in a number of ways, and there are many books devoted to the subject, as well as PR companies and consultants out there wanting to help you. The routes you choose will depend upon the business you are aiming for (e.g. local, national or global) and the markets you want to serve. The key is to connect with the right company at the right time – when they are looking for a molded parts supplier, and specifically one with the capabilities and expertise that you have to offer.

Traditional routes include advertising in printed magazines and making sure you are listed in all the relevant directories. Since the advent of the internet there are also digital magazines to consider for advertisements, which can reach a very wide audience and also include direct links to your website. There are also many more online directories and buyers' guides. Editorial copy in magazines or even local newspapers can be more valuable, so make sure you send out announcements to the press if you are investing in new plant for example. Networking and personal contact can work well through attendance at trade shows, conferences and conventions with a booth or a table-top exhibit, and getting involved with trade associations. If you are a natural communicator you are probably already looking at the options opened up by social networking, such as participating in discussion groups and blogging to raise your profile.

Differentiation is key to your marketing method, no matter which of the above methods you choose or if you choose to do all of them. What sets you apart from the crowd? The days of being all things to all people are long over, even if you are genuinely a 'custom' molder for a diverse customer base in diverse markets. Marketing methods can be difficult to develop until you understand your differentiating factors. For example, most brochures look similar and list capabilities such as machinery type and size, software, mold design, engineering and moldmaking. But most custom molders have all of these, so it's more important to promote those things – machinery, software, expertise, etc. – that set you apart. And it's much easier now to provide your customers with this additional information. You are no longer constrained by the cost and format of the traditional glossy brochure. Most things are done online now, including brochures, newsletters, promotional pieces, press releases and educational articles. So here are some keys to promoting your differentiating factors:

1. Begin your promotional materials with those factors that you have determined are your forte; those things you do best and those markets that you serve successfully and profitably. For example: "*The aerospace industry demands high-temperature engineering grade plastics, and not every molder has the expertise, the talent and the technology to mold these materials, such as Ultem, PEEK, Amodel and others. For over 20 years, XXX Molding Company has been a supplier of high-temperature components made from engineering grade materials to the top aircraft engine OEMs. If your components are for high-temperature environments or are subject to chemicals, oils, gasoline and other corrosive environments, XXX Molding Company has the solution to your challenges.*"

2. Let your customers speak for you. Get quotes from your current customers, and if you win awards, or receive congratulatory letters for a recent job well-done, put those on your website where they are readily seen. Current, satisfied customers can be your best sales people.

3. Potential customers typically look at website homepage for just a few seconds so first impressions and a professional appearance are important. To keep them looking longer, and to make them want to look further, consider making your website active and interactive. People like to watch 'movies' so provide some action. Molding machines with robotics showing the technology of high-speed molding and extraction; secondary operations in action; high-speed machining with lots of machining fluid flying; EDM machines sparking. These are all ways to capture a viewer's attention, and *SHOW* your technology, not just *tell* about the technology you use. It's an old truism in the writing world, but it certainly applies here: *Show* me, don't just *tell* me!

Appendix 2.1: Employee survey questionnaire template

Part I: Who are we?

1. Who are we? Provide your definition of the company.
2. What do we do (i.e. What type of capabilities, services, etc. do we provide to our customers?) – List 3 or 4 of our primary capabilities.
3. What do we do best, i.e. what are our greatest strengths? – List 3.
4. What do we need to do better, i.e. what are our weaknesses? – List 3.
5. On a scale of 1 to 10, with 10 being excellent, where would you rate our company in terms of:
   a. Overall quality of molds (produced in-house or outside)
   b. Overall quality of our engineering capabilities/services
   c. Overall quality of molded parts produced
   d. Communications, both internal (among team members) and external
   e. Teamwork
   f. Relationship building with customers
   g. Response to customer concerns, inquiries, etc.
6. How would you compare our company with other custom injection molders that you've read about or heard about?
   a. In the top 10% of the best companies
   b. In the top 20% of the best companies
   c. An average company
   d. Below average company

Part II: Who are our customers?

7. What is the 'ideal' customer for our company? (Give 3 or 4 attributes of that customer)
8. Name three of our best customers and one reason you rate them as such.
9. What does our company need to do better to increase business with these 'best' customers and grow the company?
10. What capabilities do we need to add to achieve our growth objectives with current, satisfied customers?

11. Name one short-term goal you would like to see the company accomplish in the next 90 days that would improve our relationships with current, satisfied customers.
12. If we have lost business from current customers over the past year, why did we lose that work? Give 1 or 2 main ideas.
13. How could we have kept that work (assuming it was business we wanted)? Give 1 or 2 ideas.

Part III: Potential customers/markets

14. On average how many RFQs do we receive monthly from current customers?
15. On average how many RFQs do we received monthly from new, potential customers?
16. On average what percentage of RFQs that we quote turn into jobs?
17. On the quotes we lost to a competitor, list the most common reason(s) given:
18. In your experience, what is the most effective way to obtain new customers or enter new markets? (direct sales, advertising, better website, etc.)
19. Rate the effectiveness of the following methods of promoting the company to gain new business (10 = most effective, 1 = least effective):
    - *a. Direct sales people (working only for our company)*
    - *b. Independent sales reps*
    - *c. Internet/website*
    - *d. Print advertising (magazines, trade newsletters, etc.)*
    - *e. Press releases*
    - *f. Joining industry/trade associations, networking*
    - *g. Word of mouth from current, satisfied customers*
20. What is one short-term goal (next 90 days) that you would like to see us achieve in the area of marketing and sales?
21. What is one long-term goal (in the next year) that you would like the company achieve in the area of marketing and sales?

Part IV: Who are our competitors?

22. Name two custom injection molding firms that you believe are our top competition:
23. Why do you think they are so competitive to us? Capabilities? Price? Other?

24. Name three things that you believe we need to do to be more competitive with those companies to which we often lose business:

Please feel free to make additional comments, suggestions, etc.

Appendix 2.2: Customer survey questionnaire template

Ideally your sales team can ask these questions when they are out visiting a customer, and have his undivided attention. If you have to complete the survey over the phone then make sure you are talking to the person who knows your company best, and agree a convenient time for him. The following questions should only take a few minutes to answer, and should improve the service you can offer him. They can also be tailored to reflect your business more accurately (omitting the last two for instance, if you are not in the business of moldmaking).

1. What do you like best about doing business with [say company name]?
2. What do you like least about doing business with [say company name]?
3. How would you rate the following aspects of the service offered by this company on a scale of 1 to 10, with 10 being excellent and 1 being less than satisfactory:
   a. *Design innovation/creativity*
   b. *Quality of molded parts*
   c. *Quality of molds*
   d. *On-time delivery*
   e. *Engineering assistance*
   f. *Customer service*
   g. *Sales [Do our sales people contact you often enough? Too much?]*
4. If there were one area in which we could improve that would be the most beneficial to you and your business, what would that be and why?
5. In your opinion, from the following list what will be the most significant factor driving change in your business over the next three years?
   a. *Expanded markets (national and global)*
   b. *New product introductions*
   c. *Increased production in existing plants*
   d. *Increased or added production globally*

6. On a scale of 1 to 10, with 10 being very important, how important is the idea of a strategic alliance with your suppliers to your company?
7. What added or expanded capabilities would you like to see us acquire in the near future to keep pace with your company's projected growth and manufacturing requirements. [Offer ideas such as greater product design assistance, mold tryouts/qualification, prototyping, and also 'finishing' techniques such as hot-stamping, assembly, packaging, etc.]
8. Currently, where does [say company name] rank in your supplier base on a scale of 1 to 10, with 10 being at the top as one of your most valued suppliers, and why?
9. What is the most critical factor in your selection of mold suppliers?
10. Today's mold suppliers are striving to be solutions providers, offering engineering services, secondary operations and value-added services. On a scale of 1 to 10, with 10 being very important, how important is this drive toward being a 'full-service' or value-added supplier to your company.

Please feel free to make additional comments, suggestions, etc

Chapter 3

Marketing communications – telling your story and communicating your value

Once you've identified your company's strengths, capabilities, ideal markets and customers in those markets, your next step is to get the word out to your customers and potential customers. You have to tell your story and communicate your value so that you will attract those customers you want in order to grow your business.

> ☞ *"Many manufacturers fail to devote the quality time that it takes to make their marketing effective because many still look at marketing as a luxury item".*

Greg Hanoosh, founder of Next Step Communications, a marketing communications firm specializing in the plastics and packaging industries, said, *"Manufacturers who engage in marketing should expect to increase their name recognition; they should expect to generate leads. And, they should expect a return on their investment, but this is easier said than done. The truth is, no matter what you may hear, much of what we call marketing communications is not an exact science and results are not usually black and white."* (**www.next-step.com**).

Bob Martino, founder of Martino Communications, a public relations agency serving clients in plastics since 1993, once said that **'build it and they will come' does not apply to running a business**. It takes a lot of resources to develop a marketing and communications strategy, so you need to develop a strategy that maximizes your investment (**www.martinocommunications.com**).

Often small business owners such as many of those in the molding and moldmaking industries can't see the value in the intangibles of marketing and marketing communications. Certainly, it's easier for those who are focused on the tangibles of 'machinery' to see the value in a new CNC machining center or a new 300-ton injection molding press.

Nanette Gregory of J4 Communications, another marketing communications firm with a plastics industry focus, said that lack of commitment from molders and moldmakers can come from how marketing is perceived within a business framework. *"Many manufacturers fail to devote the quality time that it takes to make their marketing effective because many still look at marketing as a luxury item"* (**j4publicrelations.com**).

I've often stated that for many molders and moldmakers, if they can't kick it and put oil into it, they can't see the value in it. In other words, they can see the value in a new molding press, new design software, or a new CNC machine – all the 'hard' items in the world of manufacturing. But when it comes to the 'soft' items – those things that are less tangible than equipment – it becomes more difficult to quantify their value and qualify their role in a good business model.

B2B (business-to-business) marketing is more difficult in some ways than marketing to consumers. It takes work to develop your marketing strategy (Chapter 2) and then implement a communications plan that proactively promotes your company in the markets and with those customers you have targeted as your best opportunities for new business. Your communications plan must be tied to your marketing strategy or your investment in communications might be wasted.

Whether the marketing communications plan consists of a formal advertising campaign, public relations efforts, trade shows, direct mail, email, a website, or a mix of some or all of those, **having a consistent message and sticking to that message, is key to its success**. Greg Hanoosh has commented, *"I've seen some companies go full tilt with a marketing communications plan one year and virtually disappear the next. This wastes a lot of the momentum and image gained from a campaign, and when companies start up again, it's like starting from scratch."*

Sometimes companies literally disappear off the face of the earth because they stop sending out press releases or news items, or advertising or even sending e-mails. And some of these are large molding companies that are well known and at the top of the industry. Perhaps they believe that since they are so well-known, they don't have to make any efforts to tell their story or promote their company; that customers will flock to them without any effort at all. However, given today's global competitive environment, no company can ever assume that it's too big to promote itself any longer. You need to remain visible to remind your customers that you exist. Remember the Aesop fable of the Hare and the Tortoise! Stopping to rest means others have the opportunity to overtake you in the marketplace.

Marketing communication tips: Scheibel Halaska

Mary Scheibel, principal of Scheibel Halaska, a PR and marketing communications firm identified the following five key steps to getting started on a strategic marketing initiative.[1] The goal is not only reacting to marketplace

1 Creating a processor identity, August 31, **2004**, www.plasticstoday.com.

changes, but to ultimately anticipate those changes to stay ahead of the curve – and the competition. (**www.insidesh.com**).

Start with a Business Strategy. Know your value proposition and know who in the marketplace will value what you offer. A good strategic plan should articulate a go-to-market strategy that includes an understanding of what your customers currently value, your capabilities, distribution channels and revenue/profitability goals.

Create a brand. Your brand is your promise. It's what you want people to think about when they think about your company. We all recognize the value of consumer brands, yet few business-to-business contract manufacturers are creating brands of their own.

Integrate internal and external communications. Great brands are created inside organizations by employees who fulfill on the company's brand promise every business day. Strategic thinking can't be left in the boardroom. The role of brand-based marketing communications is to explicitly declare your commitment. This makes it critical to invest in communications that educate and mobilize employees. People accept your brand story or not based on the experience they have encountering your people, processes, products and services.

Execute creatively. Great creative marketing communications initiatives pack more than just stopping power. They captivate people and influence their actions. While strategic thinking is critical, it falls short of the goal unless strategic thinking is integrated with creative execution. Contract manufacturers are notorious for communications that simply describe their services and feature poor quality photographs of building fronts and manufacturing equipment that could be anywhere. Creative marketing communications that do not differentiate your company damage the integrity of your brand and diminish the value of your investment.

Get professional help. While most manufacturers wouldn't dream of having anyone less than an expert toolmaker design a tool, most rely on insiders to design their marketing programs. The reality is that strategic marketing simply isn't a 'core' competency in most companies. Plus, it's hard to see the front of the house from inside the house. By bringing in outside counsel, you bring in new ideas and perspectives that aren't weighted by power, position or authority in the company.

3.1 Branding isn't just for soft drinks

While we might think of branding as something that Coca-Cola does with its soft drinks, or Procter & Gamble does for its health and beauty products, the first thing we need to learn is that **branding is for every company**! You don't have to be a large, publically traded OEM with lots of proprietary products that are sold to retailers everywhere, to have a brand.

> ☞ ***"If you don't have some type of brand or compelling story about your unique value proposition, you're leveling the playing field. Your customers will think all molders or moldmakers are the same, so all you have to compete on is price."***

So what is branding? **Branding involves differentiating your company from the dozens or even hundreds of competitors out there that do basically the same thing you do in your same space (i.e. in the same markets and with the same customers) via developing a unique identity**. It means getting your brand so ingrained in the minds of your customers and potential customers that they will think only of your company whenever someone says 'industrial molding' or 'unscrewing molds' or 'under-hood component molding.'

It takes time and a lot of work to first identify your brand, which you do at the front end with your strategic marketing plan, and then to begin proactive steps to promote that brand. For example, when someone asks about a company that can make very large, complex multi-cavity injection molds for the medical and pharmaceutical markets, one of the first companies likely to come to mind is Tech Mold Inc. (**www.techmold.com**). While that company makes other molds as well, nearly all of their work is in the area of very large, multi-cavity molds, with their primary customers being those in the medical and pharmaceutical markets. Tech Mold clearly has a brand. They actively promote their capabilities to those two primary markets, but frequently also find work in other markets that have similar requirements for large, complex multi-cavity molds.

According to Mary Scheibel of Scheibel Halaska, "*Your brand reputation can grow with you if you position your company correctly.*" Creating brand awareness among your customers and potential customers, and positioning your company in the marketplace are similar, but serve different purposes. "*Positioning your brand in the marketplace answers the question, 'What is the space I want to occupy in the minds of my target audience?' Branding is about differentiating. If you don't have some type of brand or compelling story about your unique value proposition, you're leveling the playing field. Your customers will think all molders or moldmakers are the same, so all you have to compete on is price.*"

And therein lies a huge problem for molders and moldmakers in today's competitive environment. Too many purchasing agents, engineers and others tasked with buying molds and molded parts fail to understand the value and never see the differentiating factors among molding or moldmaking companies.

Because the value is not understood (arguably due to communications failure on the part of the molders or moldmakers) everyone is left competing on price. And with the Chinese molding industry thrown into the mix, this is driving everyone to the lowest common denominator.

> ☞ **One molding facility pretty much looks like the next!**

Another challenge for molders and moldmakers seeking to identify their brand and position themselves in their target markets, is that few of them have unique processes or machinery to set them apart from the crowd. An injection molding company is an injection molding company; each one has injection molding presses, probably moldmaking or mold maintenance/repair capabilities with CNC machining centers and mills, they have a quality lab with necessary inspection equipment, and they have materials handling equipment. One molding facility pretty much looks like the next!

While some molders specialize in very large molds for large parts, and may have presses over 500 tons clamping force, whereas others might specialize in micromolding in very small (under 40 tons) presses, there will still be many competitors in each space. There are many molders who do many of the same types of work for a diverse customer base, and claim to be all things to all people. Just about all molders claim to have 'on time delivery' and produce 'precision complex parts' in 'state-of-the-art' molding operations. So where's the uniqueness? The differentiating factor?

That means there must be something else – some intangible uniqueness or differentiating factor that separates each molding company from its competitors. If this is absent then there's not much chance of staying in business for the long term. In fact, **many molding and moldmaking companies have gone out of business during the first 10 years of this millennium just because they lacked these differentiating factors. They couldn't stand out among all the competition, so they faded into the background and fell out of the race**.

For example, you can be more than just a molder by adding a unique capability that not every molder has, such as providing up-front product design assistance through rapid prototyping services (see Chapter 14 for more detail), or adding more secondary operations. You can differentiate your company by putting yourself in a niche such as a molder of very small parts, or even micromolding which is a relatively new niche.

> ☞ **You can be more than just a molder by adding a unique capability that not every molder has.**

☞ **Finding a niche may not require any changes from what you've already been doing. Sometimes it's just a case of recognizing that you've become really good at doing a particular thing, and taking that to the next level.**

Donnelly Custom Manufacturing Company (**www.donnmfg.com**) is an example of a company with a well-defined brand. Since its founding in 1984, the company had seemed to buck the trend by doing a lot of short run work. While most molders dream of the long-running jobs of millions of parts, Donnelly chose to focus on doing short runs, using the tag line 'How Short Run is Done' to define the company's niche and put Donnelly front and center for these jobs. The company promotes its ability to provide compressed lead times, fast mold change-overs, and global supply chain management.

Finding a niche may not require any changes from what you've already been doing. Sometimes it's just a case of recognizing that you've become really good at doing a particular thing, and taking that to the next level.

3.2 Communicating the right message

The right message is the one that shows how your company differs from all the others that occupy the same space, and one that is conveyed in an effective manner so that people actually 'get it.'

Case history: C. Brewer Co.

Several years ago, Chuck Brewer III, CEO and the third generation to run C. Brewer Co. (**www.cbrewer.com**), began taking a look at the company, not only from a moldmaker and molder's viewpoint, but from a marketing angle. One thing he knew was that marketing a custom molding business doesn't lend itself to well-defined areas of a marketing campaign.

Brewer joined an executive mentoring organization that provided one-on-one executive coaching. One of the benefits that Brewer received was an understanding that he needed to be more aggressive with his marketing. He discovered some important keys to developing a marketing strategy that led him to a higher level of promoting C. Brewer Co. *"It's more than just marketing, it's marketing with the intent to inform,"* said Brewer. Armed with the information gathered from doing extensive customer surveys (see Chapter 2 for ideas on this), the company was able to take a fresh look at itself through the eyes of the customer. *"We saw who we are and who we're not, then discerned for ourselves whether that was an accurate perception or not".*

That is a key point. Perception is reality! Regardless of whether you believe that your customers' perceptions about your company are mistaken, it is their reality. Now it becomes your job to inform your customers and potential customers about your company and who you are, and educate them about how you can provide solutions to their problems. Educating and informing should be the number one goal of your marketing communications efforts.

Brewer created a tag line that he felt told customers what C. Brewer Co. was all about, 'Perfecting the Science of Molding,' and then began sponsoring educational opportunities for customers such as its online 'eMovies' to educate them about scientific molding. He has also produced a series of white papers on time, temperature, pressure and cooling to help educate customers on the company's approach to molding technology.

C. Brewer's strategic marketing planning gave Brewer the knowledge he needed to position C. Brewer Co. in the market through an effective marketing communications strategy. *"It helps us promote what we have to offer that other molders do not,"* he said. *"We do this visually and with the copy we write. We try to create total visual unity of C. Brewer, with all of our promotional material - print image and on the website."*

This total visual unity highlights another aspect of communicating the right message; Effective marketing communications are also about putting out a *consistent* message over time and throughout your company. **Many companies have found that a clear mission statement helps to create and maintain this consistency**.

3.3 Websites

In the 21st century, designing and maintaining an effective website is another essential part of your marketing activity.

Todd Schuett of Creative Technology (**www.creat.com**), who develop websites for customers in the plastics industry, points out that *"Your website is a 24/7 ad for what your company does."*

The first thing that is needed for creating a website or updating your current site is to analyze your company and customers, starting with a SWOT evaluation (Strengths, Weaknesses, Opportunities and Threats). *"The key elements to feature in your website are your strengths,"* according to Schuett. *"Based on your evaluation of your company, what is your specialty? Promotional materials including the website should be designed to your strengths and speak to those markets and customers you have identified through your evaluation and marketing strategy, where you want to capture business."*

Your website should also highlight niches and areas of specialization. Perhaps you have a unique niche, such as the ability to make threaded parts. Then show threaded cores and how they work to show what you do. Potential weaknesses can also sometimes be turned around – perhaps you have a concern that like many molders you are only a small business. This may indeed limit your ability to take on certain jobs, but conversely remaining small may help you provide a flexible and fast service.

Schuett suggested that when updating your current website, you should ask yourself a few questions:

- Where are we right now? Do I like my site? Is it up to date? Does it accurately depict what we do? Is it visually appealing?
- In order to move forward, think about who your website is for. Internal customers? External customers? Vendors? Employees?
- Is your site engaging? To make your site more engaging, tell your story with photos. Text is important, but 'showing' your capabilities is better than just telling. Engaging visitors in your site and getting them interested long enough to stay and learn more about who you are is the first goal.
- Next, you must know who your website is engaging. A lot of different of people look at your website, however your main targets are likely to be engineers and purchasing people. While engineers generally understand moldmaking, molds and molding, buyers may not understand what they're buying. Schuett suggested that your site be an educational tool for those who may not know exactly what a mold is or how it works.

3.3.1 Priorities to show your audience

Show them what you make: have photos of plastic parts and mold examples to convince prospects that you can do what they need done. How you portray that is also important. Even if you are a moldmaker and not a molder, pictures of excellent complex moldings can remind customers that quality molds are crucial for quality products.

Business development: Show that your company is growing, and include things such as news items or articles from industry trade magazines that feature your company.

Show, don't just tell: Show how you make your tooling, i.e. use video to show your capabilities. Show people in action. Also, don't just show a plastic nozzle, but show and describe a very complex, precision nozzle. If you make a mold for something that people purchase and use, put a photo of that part on your site so people can relate to it. Showing parts to which everyone can relate shows that you are part of their life and have an impact through the molds or parts that you make.

Build your photo portfolio: Remember to take photos of molds and molding action as you go so you have plenty of promotional shots. When you finish a mold take a photo of it. When you do first article sampling and the parts come out looking beautiful, take a photo and build a file. That way you'll have plenty of photos for your website.

Show detail on the photos: Close-up shots tell a lot without necessarily telling it all. Fill the frame with a core or cavity or a specific complex area of a part, to demonstrate your particular skill to the experts in your audience.

Site organization and content are critical to the site's appeal and the customers' experience, and many books and articles have been devoted to the subject. You may want to look at lots of other sites, and decide which features you find useful and attractive. There are no hard and fast rules, but there are certain conventions and expectations of what visitors expect to find on a home page for instance.

- Consider how you want the customer to interact with the site. They will certainly need a means of contacting you, but do you also need a secure e-commerce area? You may want a secure FTP facility for customers to upload their CAD files for example. Can you offer a benefit if visitors register on your site and provide their contact information? This could be access to downloadable white papers or tutorials.

- Think about how you will keep your site up to date. Can you make any necessary content changes quickly and easily?

- Does a large proportion of your work come from other countries, and would it be helpful to translate the site into different languages?

- Use your website to introduce key personnel and put faces to names. Consider including some historical information if your company has an impressive pedigree.

Todd Schuett also made some more suggestions:

- 'Careers' can be an important tab. Listing job opportunities is not necessarily just a call for people to come and apply for jobs, but a signal that you are growing and investing in the future. A training section shows you're re-investing in your growth.

- Are you involved in politics and out lobbying for the interests of your industry? This is good to include as it shows you're actively trying to make a difference in the world. It shows community involvement. On one site, the company shows employees helping to build a house for Habitat for Humanity. It shows that you're trying to make the world a better place.

- Having a photo of the facility isn't all bad. A nice looking facility with some green grass and blue sky is important, but not the biggest thing. We see less and less websites showing buildings and delivery trucks. On the other hand, it can show that you have a modern and clean facility. It can also be helpful to have a map to your facility.

The point is that there is no ultimate site and you never get to the perfect site permanently. "*We all look at things differently, and things at our companies are constantly changing,*" Schuett noted. "*Tell what you do quickly because you have just seconds to lock them in or you'll lose them. Make everything count on your site: Text and photos/illustrations. Build some things in so they have to hunt for it but give others the fast overview quickly and easily. Appeal to young and old alike – attract the young into the trade with photos of youth, and make your site photographically appealing with things like motorcycles. Text is essential not only for descriptions but for search engines, so make sure you have good text.*"

One thing you could consider including is video. It helps you not just tell your story but SHOW your story. According to Schuett; "*It has to be good but it can be a real bonus, and help portray your culture and what you do.*" However, while your site has to be visually appealing, you should avoid too many distractions such as background music or useless animation which interfere with the real message. (Elaborate sites with lots of illustrations can still be frustratingly slow to load.)

Remember that ongoing updates are critical. "*It can be the most important thing you do to keep it relevant, current, and interesting to visitors to the site,*" concluded Schuett.

You should always give customers the opportunity to contact you via your site at any point throughout their exploration. Every page should have the 'Contact Us' tab clearly visible. Additionally, make your site consistent with all of your marketing materials. It should have the look of your brand – that unique identity that you've designed for yourself based on your evaluation of your strengths. Focus on those throughout your promotional material, including your website, your brochures and other company literature, your trade show booth, whatever you put out there to promote your company.

Finally, you will benefit from at least a little insight into the technology you are dealing with. You may not have the time or interest to research this yourself in detail, but there are plenty of people who can help you with the following types of issues. As we all use the internet increasingly in our home lives, remember that users expect the same convenience and useability from your site as you would when you order your groceries, book a holiday, research your ancestry or perform any other task online.

- Search engine optimization (making your site as visible as possible when potential customers search the internet) is a science in itself, and a moving target. Read up around the subject, or employ someone who understands it to build or maintain your site. Relevant keywords and links from other sites are a good starting point here, so consider providing some useful tools or information to encourage users to recommend and return to your site.

- Be aware of the rapid changes in technology. A few years ago you could expect your customer to be looking at your site from a desktop PC, with a printer close at hand. Now he is as likely to be browsing from a tablet or even his mobile phone. All these different platforms require different functionality to perform best. Typically smaller screens will benefit from less text for example.

An important point here is the speed of change, and the fact that your site must evolve with it. You will need to invest time and money here on a regular basis to get the best return.

3.3.2 Internet marketing – more than just a website

As the internet has changed the way many people live their personal lives, so it also impacts the ways they promote their businesses and interact with new and existing clients. **Social networking has reached the business world**, with many companies interacting through interest groups on sites such as LinkedIn. Others regularly post articles or even present webinars on the various industry platforms, communicate through message boards or supply articles to populate knowledge bases. Personal recommendations and networking have always been important in business, but these activities are now also duplicated on the internet, with the ability to share your views, likes and dislikes with colleagues and 'followers'.

Arguably these typically small-scale promotional activities have brought back a greater element of personal contact to business marketing, whilst allowing these contacts to be located anywhere in the world. You can also gain some useful insights into the topics that are concerning or exciting your colleagues and competitors. However it cannot be disputed that they are time-consuming, and need to be managed and evaluated in the same way as any other marketing actions, Try signing up to some interesting groups before deciding whether this is another way forward for you.

3.4 Publicity tips: how to get noticed in the press

> ☞ **One thing's for certain: if you never send out any publicity about your company, you will never get noticed!**

Getting publicity is not easy. It takes persistence and perseverance – **the squeaky wheel definitely gets the grease in the publicity game.** But one thing's for certain: if you never send out any publicity about your company, you will never get noticed! Many smaller company owners in particular have a difficult time figuring out how to get noticed in a sea of processors, moldmakers, and suppliers, and the bigger companies that often seem to garner all the attention. That's because the larger companies tend to have public relations (PR) or media relations people on staff whose job it is to get publicity out to the media as often as they can.

Granted, some seem to send out a press release every time someone in the company trips on the carpeting. There is such a thing as flooding the media outlets with so much trivia that they no longer pay any attention to you. Understanding what is news and then providing the 'hook' for that news item to the media will get you noticed. Buying new equipment, renovating the plant, expanding the plant or siting a plant in a new location is big news, especially during times of economic downturn and bankruptcies. Of course bankruptcies are news as well, and often PR involves being proactive about bad news or stopping rumors about your company.

Many companies that can't afford a full-time media relations person will hire a public relations firm to which they pay a retainer for their services for a certain period of time. They are professionals and can help you determine what is 'news' and how to disseminate it to the media; they might write case studies to educate your customers and provide good fodder for feature articles in the media; and generally spot opportunities to get your company noticed.

A few tips for getting noticed include:

Know your message. Who are you? What message do you want to send? Who do you want to see your message and how do you want them to respond? What market niche are you after? You have to be able to answer these questions or else you'll just be firing words out to everyone and being relevant to no one.

Have a focus. Many times the real story is hidden. Instead of sending a press release about receiving your ISO 9001-2008 certification (which is not news nowadays), grab a reporter's interest with some relevant details. What prompted you to get certified? Did you just land a big job with a large, multi-national OEM and they required it as a condition of getting the work? What management stresses did it cause? How did you resolve them? Or, don't just put out a press release saying that you just bought a 500-ton molding press. So what? Tell the story! Why did you buy a 500-ton press? It can be very big news if it's the largest press you have on the floor. Did you just land a nice program in which one part required a 500-ton press? Will you be molding a new product? Give the reporter a true hook on which to hang a really good article. **Otherwise, your press release gets relegated to the 'slush' pile – and we all know what happens to slush!**

Let your customers (or your suppliers) do the talking. We all know what you think of your company. A better way to promote your company and get publicity is to let your customers (or your suppliers) do the talking. Did you solve a particularly difficult design problem for a customer? What did the customer say about your ability to succeed where others failed? Ask your customers if they will comment about your expertise and capabilities as a supplier to them. And don't forget about your suppliers! Are you a good customer for a particular brand of machine tool or processing equipment? Ask your supplier if they need a spokesperson for their next ad campaign. Provide them with a case study on how you use their machinery, equipment or resin to increase productivity, accuracy in machining, and profitability. They know the same rule: let your customers do the talking.

How to get noticed in the trade press

With upwards of 200 e-mails per day coming to the computer of editors of trade magazines, getting noticed might seem like an impossible task. But, small things can make a big difference. These sometimes seemingly insignificant items often are deciding factors whether press information from a molder or moldmaker actually make it into the print publication or online. To ensure better chances, your news has to be eye-catching, brief and to the point, and more importantly, put the 'hook' that you want editors to grab, right up front. Here are five tips for getting noticed in the trade press:

Send good photos with the release. The emphasis should be on good quality and pertinent to the news. Good photos are 'high-resolution' (generally 300 dots per square inch – dpi), which are required for good trade magazine reproduction. If it's people news such as a new president being named to the company, then include a good 'head shot' photo of this person. Always include the caption with the .jpg or .tif photo – hundreds of photos land in the in-boxes of editors so make sure he or she can recognize what the photo is and who it's from. Be sure you've spelled peoples' names correctly!

Provide contact information. Every press release should have good contact information including the name, phone number and e-mail of the contact on the press release itself, not just the e-mail. This makes it easy to contact the right person for any additional information, questions or comments needed.

Provide contacts that can answer both corporate and technical questions. Many times a technical question comes up but no one is available – or the only person allowed to speak to the press is the president of the company. It helps to have all your bases covered so the story can be complete.

Know the deadlines for editorial for any given month. Many people incorrectly assume that magazines are produced only days before publication. While online publications can be produced overnight, print magazines need anywhere from two to six weeks to prepare the magazine for printing. So, if there's specific issue you want to get into, say a medical special feature for a particular month, check the editorial calendar on the magazine's website so you'll get your material in under the deadline.

Provide good case studies. Case studies, particularly where technology news is concerned, really help describe the benefits and/or the advancements of the technology. A mold design that offered a new and different molding solution for a customer, for example, would be an excellent case study. Try to involve the customer, and if you are going to use the customer's name, get their permission first. Some OEMs don't like to make public what they are doing with regard to manufacturing. However, while an 'anonymous' customer isn't the best, it's better than no case study at all.

3.4.1 Trade publications

There are many types of trade publications. Traditional printed magazines almost all have a digital counterpart now, but there are also wholly digital magazines, such as AMI's Injection World. The lower production costs associated with electronic-only production enable some of these newer titles to target much smaller market sectors. There are also electronic newsletters, communities, noticeboards and similar information services distributed weekly or even daily, many of which consume and recirculate press releases. Trade publishing has changed rapidly with the growth of the internet, so it is important to keep your contacts up-to-date, to ensure you are making the most of all your opportunities.

Trade magazines targeted specifically at the plastics industry include the following:

Plastics News (**www.plasticsnews.com**)

European Plastics News (**www.europeanplasticsnews.com**)

Plastics & Rubber Weekly (**www.PRW.com**)

Injection World (**www.injectionworld.com**)

Plastics Technology (**www.ptonline.com**)

Moldmaking Technology (**www.moldmakingtechnology.com**)

Plastics Decorating Magazine (**www.plasticsdecorating.com**)

There are also manufacturing and design-related magazines which contain editorial about plastics applications:

Design 2 Part (**www.d2p.com**)

Design News (**www.designnews.com**)

Manufacturing Today (**www.businessmediapublications.com**)

Quality Magazine (**www.bnpmedia.com**)

There are many other trade magazines serving the markets which plastics companies want to reach, for example medical devices and packaging:

Medical Device and Diagnostics Industry (**www.mddionline.com**)

Medical Product Manufacturing News. **www.qmed.com/mpmn**

Medical Design (**www.medicaldesign.com**)

Medical Product Outsourcing (**www.mpo-mag.com**)

European Medical device Technology (**www.emdt.co.uk**)

Packaging News (**www.packagingnews.co.uk**)

Packaging Today (**www.packagingtoday.co.uk**)

PackagingEurope (**www.packagingeurope.com**)

Plastics in Packaging (**www.plasticsinpackaging.com**)

Cosmetic & Personal Care Packaging (**www.cpcpkg.com**)

Pharmaceutical & Medical Packaging News (**www.pmpnews.com**)

Packaging World (**www.packworld.com**)

There are also trade associations that publish their own magazines to their members such as:

The Society of Plastics Engineers – **www.4spe.org**

The Society of Manufacturing Engineers – **www.sem.org**

The Association for Manufacturing Excellence – **www.ame.org**

Online communities and news services covering plastics include the following examples:

www.azom.com

www.ides.com

www.omnexus.com

www.peplus.co.uk

www.performance-materials.net

www.netcomposites.com

www.plasticstoday.com

And no matter what markets you serve, there are specialist magazines that also serve that market, where you can send information, case studies, articles, and news, whether it's automotive, lawn and garden, heavy equipment, consumer electronics, and more. They can be accessed online so just Google the markets you want to find trade associations, magazines and other opportunities to reach the businesses you need to reach.

Most magazines will also take paid advertising, and you may wish to consider this for a new product or service. It can be cheaper than you imagine – particularly in the digital magazines, where you can also have a 'clickthrough' to your website or a pdf version of a brochure or catalogue.

3.5 Trade shows and industry conferences

Trade shows provide a great way of meeting customers and suppliers, seeing the latest materials, machinery and product developments, watching the competition and generally getting a feel for the mood and direction of the

industry. 'Walking the aisles' can provide an excellent introduction for someone coming new to the plastics industry, or a refresher for the expert who rarely escapes from the factory floor. If you plan ahead you can also use your time at a show productively, to make appointments with many of your key contacts while they are all in one place. Many trade shows also have a free lecture program running in parallel, as well as other satellite events – so check out how to gain the most intelligence and new business from your time.

Industry conferences can also provide a great opportunity to network with a whole lot of industry contacts in a short time, whilst also typically learning about the latest technological developments or issues affecting your particular industry sectors. Concerns about environmental issues or health and safety, which could impact your choice of materials for example, are frequently discussed in conference programs and around the lobby.

Trade shows and conferences are also useful for evaluating potential new markets, and understanding how a sector really works before deciding to change the direction of your business and make substantial investments. For example, if you think you really want to break into medical molding, a look at the exhibitors at a show will tell you a lot about the business. Which materials suppliers are present? What are the machinery suppliers promoting? What sort of ancillary equipment is on display that you might require? Who will you be competing against? You can also gauge whether a sector is generally upbeat or struggling. Visiting a national or regional show could be a good way of looking at a new country and understanding how the business works there.

Some of the trade shows in particular are huge, occupying a dozen or more halls. Travelling to these shows, particularly as an exhibitor can appear prohibitively expensive. However, there are frequently government subsidies available (perhaps to encourage small businesses to explore new export markets), so take a look at this type of option before dismissing the idea. Trade associations sometimes administer these schemes, and may also offer the chance to travel and exhibit jointly with other companies, minimising costs and administrative hassle for you.

There are also much smaller local events, which can also be effective for developing business if many of your potential customers are relatively close to you. Sometimes local business development agencies promote regional business by running industry 'matchmaking' events (meet-the-buyer and similar formats). These may be focused on particular end-use sectors, and enable companies to take a small stand and then have a whole program of pre-arranged sales meetings with potential customers brought in from that industry sector. Look out for these events in your area.

There are dozens of trade shows each year, some specifically for the plastics industry, others more general manufacturing or design, and many others geared toward specific industries. There are national and international shows, and many regional ones. You can search the web to find these or many of the magazines have events listings. One of the best for events designed for the plastics industries and industries heavily involved in plastics appears on the Plastics News website (**www.plasticsnews.com**). For a broader overview of end-user industries, look at general trade show websites such as **www.tsnn.com.**The following is just a small list of some of the main shows you might want to check out:

NPE (US) National Plastics Expo, held every 3 years **(www.npe.org)**

SPE Antec (annual technical conference and exhibition) (**www.4spe.org**)

K (Kunststoffe) held every 3 years in Dusseldorf, Germany (**www.k-online.de**)

Fakuma, held in Friedrichshafen, Germany in 'non-K' years **(www.fakuma-messe.de)**

Plast, held every 3 years in Milan, Italy, Europe's 2nd largest plastics show after K **(www.plast12.org)**

Chinaplas (**www.chinaplasonline.com**)

Arabplast, held every 2 years in UAE

IranPlast, held every 2 years (**www.iranplastfair.ir**)

Pack Expo (**www.packexpo.com**)

Interpack, **held every 3 years in Dusseldorf, Germany (www.interpack.com)**

Medical Design and Manufacturing (East and West), held in the USA alongside other manufacturing events **(www.canontradeshows.com/expo)**

Design2Part manufacturing Job Shop Shows held in various US locations throughout the year **(www.d2p.com).**

Similar to the variety of trade shows on offer, conferences range from the enormous ANTEC event run by the SPE (Society of Plastics Engineers) to much more focussed meetings. For example, AMI organise around 30 conferences around the world each year, on topics as diverse as masterbatches, medical plastics, pipe fittings and electronic connectors (see **www.amiconferences.com**).

Chapter 4

Sales strategies

4.1 What do we mean by sales?

'Sales' involve the action of 'selling' the goods and services that your company has to offer. And while that sounds simple, it's actually more complex. Selling is a term that denotes 'activity' – hence, the sale is the culmination of the selling activity. Yet, the 'selling' activity begins long before the sales person sits down in front of the customer. In fact, the sale is the reward for all the upfront activities you've engaged in such as the marketing and marketing communications.

After identifying the capabilities you have, and the activities that you do best for your customers, and the customers for whom you do that best, you developed your marketing communications strategy. All of that was a precursor to the sales activity. In fact, you can't begin to sell effectively until you've done all those things.

Molders who hire a sales person and send them on the road without first providing them with a 'roadmap' are doing those people – and their company – a great disservice. They also miss the mark when they hire someone and give them the title 'Marketing and Sales Manager.' I have always believed that a molder of almost any size, but certainly a molder of $10 million in annual sales or greater, needs two people: a marketing person and a sales person.

Generally speaking, a sales person does not like to do marketing activities. A good sales person loves to be out in front of customers selling! A good sales person is one who actively develops relationships with customers. These relationships do not happen accidentally. They are nurtured and developed over a period of time. Good selling techniques involve people and personalities, and never will these personalities affect your business more than among your sales force and your customers.

Many times, customers will do business with you because they like someone at your plant. They like the company owner or the managing director or the quality manager or the sales person. The opposite is also true. In some cases a customer won't continue to do business with you because they do not like the sales person or the customer service person or the company owner. I know a mold manufacturing company owner who received a call from a customer one day, telling him, "*Don't ever send [that sales person] into my company again! He drives me crazy!*" So, the mold company owner gave the account to another sales person.

Successful sales and selling happen through strategic planning. They happen as part of a comprehensive, integrated program of marketing, marketing communications and sales, that result in your company's goals being achieved: growth and profitability.

4.2 Why you need sales staff

> ☞ **People do business with people, not companies.**

Many custom molding companies think they are too small and don't need a sales person. Many think they cannot afford a sales person. However, you can't afford *not* to have a sales person or a sales staff. You might think that your website is good enough – a sort of twenty-first century surrogate sales person – and that you can capture the business you need with that tool. But if you are in business, you need someone out there who is promoting your company and its capabilities.

People do business with people, not companies. You need that face-to-face contact in almost any selling situation, and business-to-business selling needs that personal contact as much as the retail market needs friendly people who stand inside their doors to greet customers as they walk in. And while there are people in your company that you think can fill that role alongside all the other things they do, you might find that's not true.

Sometimes molders are tempted to put their engineer into a part-time position of sales person, reasoning that since molding is a 'technical' business an engineer would be an ideal sales person. That might be true in some cases, but generally I've found that while they might be excellent engineers, they are less than stellar when it comes to selling. Personality is a huge component of selling, and first and foremost you need someone with the sales 'personality.'

Sometimes, particularly in small family-owned companies, the owner wants to perform the sales function, and if he or she is good at it, and enjoys it, and has the time to do that along with all their other responsibilities, then that's fine. But generally sales takes an inordinate amount of time: tracking quotes, following up on last week's sales calls, prospecting for new customers, touching base with current, satisfied customers, and planning next week's sales trip. A dedicated sales person who loves to sell and has that people-intuitive personality can be an extremely valuable asset.

Tim Holland, president of family-owned Metro Mold & Design Inc. (**www.metromold.com**) didn't like sales. "*Getting myself out of sales was the best thing I ever did*," he said during a panel discussion at the 2009 convention of the

American Mold Builders Association. *"I'm not good in front of the customer, and tend to back down. But the sales guy doesn't because his paycheck depends on it."*

Holland's company has a vice president of sales and five salespeople for Metro's mold manufacturing and contract injection molding businesses. It's a fairly big sales team, but Holland believes that sales people are key to his success. The firm's molding unit has 43 presses ranging from 35-650 tons.

4.2.1 Direct and indirect sales staff

Whether or not you employ a direct sales person (i.e. they work directly for your company as an employee) or an indirect (i.e. a sales rep), depends on what your company wants to accomplish. Some molding companies that want to expand their customer base geographically find that working with manufacturers' reps is beneficial. Typically these 'independent' reps serve several manufacturers of various types in a specific region. They are the consummate 'sales person' and know their territory, the OEMs and the products that they produce.

Manufacturers' reps are first and foremost 'independent' and that means they work on their own time in their own way, and they may or may not be keen to report to you on a regular basis. They generally work on commission, but some require a 'retainer' that is paid monthly to 'retain' their time and efforts while engaged in the activities of making sales calls and promoting your company.

Just because the manufacturers' sales rep is knowledgeable about manufacturing and is good at sales doesn't mean they won't need any guidance at all. In fact one of the primary reasons that a sales rep fails to be effective for a molder is that no one takes the time or makes the effort to educate him or her in the molding/moldmaking business, the types of products they are good at molding or have the press capability to mold, and the markets/customers they want to pursue. It also pays to work with the rep on a regular basis, going to their territory and meeting with prospective customers to get a feel for how the rep works, and whether or not they truly understand your business and your capabilities, products and services.

Tim Peterson, vice president of his family-owned mold manufacturing company Industrial Molds Group (**www.industrialmolds.com**), has described his solution to the direct vs. sales rep conundrum. He has one employee focused exclusively on sales, supported by the company's engineering staff, who are available to go out with the sales person to answer technical questions and collaborate with the potential customer's engineers. He also uses a sales rep specifically to cover a particular region and finds that this works for his company. However, he has commented that it takes a lot of work on Industrial Molds' part to get the best from a rep: *"I go on a regular basis to work with the rep, make sales calls with him at companies that he's tapped into, and generally keep up with the activity there."*

Hency Bunner, president of Proact 2000 LLC (**www.proact2000.com**), is a consultant working to connect the right reps with the right companies. However, even he admits that the relationship may fail without guidance and training. *"The number one key to having a successful sales rep is training,"* said Bunner. *"You've got to train them, bring them in your plant, teach them about your business and make sales calls with them."* He also recommends that manufacturers visit with the reps regularly in their territory at least four times a year, and bring them into the plant at least once or twice a year to update them on products, services, capabilities and potential customers.[1]

Putting together any type of sales force takes careful planning and selecting the right method – or combination of methods – to be effective for your company. It's also critical to have a marketing strategy in place at this point because that is the roadmap that these sales people, whether direct or reps, can follow to capture the best customers and contracts.

4.2.2 Separating inside and outside sales

Dividing of the jobs of outside sales (typically the sales manager or people hired for the express purpose of being 'outside' the plant, calling in person on buyers at OEMs) and an inside sales person (a person who remains inside and acts as the sales manager's assistant) is frequently the best way to go. An inside sales assistant to the sales manager acts as a liaison between the sales manager and the customers and potential customers. They can help to track the RFQs (is the job ready to be let? Did we get the job? If not, why not?) They can also be more available to customers when the sales manager is on the road.

In some companies the inside sales person is also responsible for customer service, which makes sense in terms of the tasks involved. Current customers are frequently the best prospects for new work. If your customer service person can handle the added work load, it might make sense to boost their salary a bit and let them handle inside sales too. It takes the pressure off of the sales manager, and provides a consistent contact point – with a real live voice – when the sales manager is out and unavailable to take calls immediately.

An inside sales person can make follow-up calls after trade shows, they can send out brochures, newsletters and other literature to current and prospective

1 Goldsberry C., Hire reps or direct salespeople? The $64,000 (or much more) question, October 13, **2009**, www.plasticstoday.com.

customers, maintain and add to your customer database, and do other sales-related tasks that the outside sales team do not have time to do. This role can be particularly effective if you don't have anyone with an exclusive marketing brief.

The inside sales person is absolutely as important as the people who do the face-to-face selling, which means the selection of this person is just as important as selecting an outside sales person. A good sales personality, detail-oriented, friendly, willing to find the answers to questions and be helpful to the prospect are just a few of the desirable qualities for this role. Also, if they are going to do cold-calling on the telephone, gather information and perhaps set up appointments for the outside sales people, this person must also be comfortable with digging for information, getting names, and approaching people they do not know. We are not all good at this. If a person has an aversion to cold calling, they won't work in this position.

4.2.3 Who sells and what do they sell?

The primary job of a sales person is to promote the company, its capabilities, services and expertise, and this is typically the easier part. The difficult part can be selling the value of the company. This means taking those differentiating factors that were established in the marketing plan and showing potential customers why they are critical to the success of his company.

The sales people also create an image of the company, ideally as one comprised of competent, skilled, dedicated people who can solve problems and provide quality throughout the process. They bring in new business by converting prospects into customers; they are generally skilled in overcoming objections that prospects might have to doing business with your company; they maintain current customers through account management and building an ongoing relationship; they also frequently act as the liaison between customers and the production floor.

Sales people act as consultants to customers such as OEMs, helping them identify specific molding challenges in a new product, then find the appropriate solution. Will a hot runner mold work best? What about a two-shot mold rather than two molds and two operations? Can this part be molded using a biodegradable material? In this industry, the custom molder's salesperson is often the person the OEM calls when it has a new product in the pipeline and needs advice. He or she may also get the call when a customer wants to make a mold modification or a material change.

To have a successful sales operation, management must be committed to the sales team by being supportive and backing them in a difficult job. Management must also define the role of a sales staff to all employees to make sure that the importance of sales to growing, sustaining, and maintaining business is understood by everyone.

Also, all the employees of the company should be aware that it's not just a salesperson's responsibility to bring in business – it's everyone's responsibility, because new business can come from unexpected places. A production manager with a custom injection molder was a having a drink after work one Friday evening when he began a conversation with the man sitting next to him. It turns out this man was a tooling engineer for a local company with tooling and molding requirements and, thanks to the efforts of the production manager, the company soon had a new contract for molding.

There was also a quality manager for a molding company who just happened to join up in a foursome at golf one morning, and another man in the foursome was the president of a large OEM in town. The two men began talking quality issues, and soon the quality manager had told this company president about the quality standards of the molding company for which he worked. A relationship developed and soon the molder had new business.

Never underestimate the power of every employee to be a salesperson, even though you might also have a formal salesperson on staff. All employees should be aware that what they do and how they do it influences the sales process for good or bad. They should work with the sales team through good communication. Many times, an OEM's engineer will make a comment to a quality person that they are thinking of moving the work if this problem doesn't get solved. That needs to be communicated to the sales staff so that they can take steps to resolve the issues. Everyone is responsible for maintaining good customer relations.

> ☞ **It's not just a salesperson's responsibility to bring in business – it's everyone's responsibility, because new business can come from unexpected places.**

4.3 Why you need a sales strategy

Does your sales team have a strategy? What tools do they use to help your business grow in the right direction with the right customers? Do they understand exactly the kind of new business you want them to bring in? Those are crucial questions to ask because selling has changed over the past decade. The presence of the internet adds some new dimensions to the sales strategy as your website becomes one of your most valued sales tools. Traditional selling –

packing up sample parts and brochures in your briefcase and traipsing all over the country cold calling no longer works. It's too expensive. It's also the 'shotgun' approach, and one of the purposes of the marketing strategy that you've completed is to provide you with a 'rifle'; to enable you to target your key markets and the key companies in those markets.

You also need a strategy and some clear targets because your potential customer can now 'visit' dozens of custom molding companies via the internet from the comfort of his office in a matter of an hour. That means your competition is far more extensive than it was a decade or so ago. A good marketing communications strategy will help you grab new potential customers via cyberspace, and provide a ready target company for your salesperson.

Lack of strategy often results in sporadic business, with a tremendous backlog one month and nothing the next. The function of a good sales strategy is to 'keep the funnel full'. A sales person once explained this as follows: "*Think of injection molding and moldmaking as a funnel, you have to keep the funnel full with potential new jobs and potential new customers, because it takes a long time for that one good job to trickle out the bottom.*"

It's true. Selling injection molding and tooling is like the old days of selling vacuum cleaners door-to-door. In this business, finding the right customers in the right markets, developing the relationship, and finally landing a job can be a long, hard-fought process. And that process requires a strategy.

The marketing strategy is the first part, with the communications plan the second part. Together these help create a sales strategy. Your marketing strategy helps you target your markets, companies and customers within those markets. The communications plan is designed to help you tell your message to those targeted markets, companies and customers. While many methods of selling require 'mass' communications such as the internet, you can still develop a 'targeted' message designed to attract those customers that you know you can serve successfully.

Target marketing/selling is targeting specific customers with your capabilities and services because your marketing plan indicates that these customers represent your best opportunity for new business and healthy growth. With target marketing you aim at specific 'targeted' customers in specific markets so that when you take a shot, you're most likely to hit the target the first time.

> ☞ **Your potential customer can now 'visit' dozens of custom molding companies via the internet from the comfort of his office in a matter of an hour.**

Custom molders and/or custom moldmakers should use a targeted marketing and sales approach because their businesses typically provide specific types of capabilities and services geared toward specific customers. **The truth is, not every OEM wants what you have to offer.** One company simply can't offer all the capabilities and services that meet the requirements of all OEMs. Those molding and/or moldmaking companies that think they can be all things to all people usually end up in bankruptcy or out of business completely.

Trying to sell to everyone means you become just one more of a large number of suppliers who are all trying to service the entire OEM community. You then become a 'me too' molder or moldmaker, which makes it more difficult to sell yourself, even if you really do believe you can supply any OEM with any requirement. Essentially, selling all things to all people usually relegates you to competing on price must like the mass merchandisers do. When you go into a major supermarket, you'll usually find all the same types of products, the same brands, the same general store layout. So, with all these mass merchandisers having no unique products, brands or identity, they end up competing for customers on price. Few molders or moldmakers can afford to offer 'blue light specials' to their customers.

Selling to a targeted market and a smaller potential customer base is much easier in some ways than trying to sell everything to everyone. With a targeted selling approach, the emphasis is placed on value – that value that your marketing strategy identified. Your target customers then become those companies that will pay for the particular qualities that only you and similar companies can offer – not just the ones who will always go for the cheapest price. This also means that you have a much clearer message to convey – because you and your customer should both understand what you have to offer, why the customer needs it, and what it is worth.

Target marketing also means that you limit your competition to those molders who target the same customers and markets as you, have basically the same type of capabilities, services and expertise as you, and can offer much of the same value. I've heard many company owners who praise their competitors that in the same league as they are, even saying, "*I don't mind competing against ABC Company because they are such a good company, very equal to the type of company we are. It's an honor to compete against them.*" That means you're competing on value and when you have reached this position you need good sales people who can sell that value to your targeted markets and customers.

To sum up why you need a sales strategy:

- Competition has intensified – your potential customer can now visit dozens of molders via the internet from his office in a matter of an hour.

- A good strategy will 'keep the funnel full', avoiding backlogs and periods of too little work.
- The 'shotgun' approach is ineffective and too expensive – a strategy provides you with the means to target the best customers more efficiently.
- With a targeted approach you can promote the 'value' of your company to those companies who need what you offer, and will be willing to pay for it.

4.4 Targeting your customers

You can think of your customer target as a real target with three rings or sections. The center section is the bullseye of the target, and this is where your current satisfied and valued customers that you identified in your marketing strategy reside. These customers will also represent the largest share of your annual sales. The 20/80 rule typically applies here, which says that 20 percent of your customers are likely to represent 80 percent of your business. Why do we put this 20 percent at the center? Because **current satisfied customers are not only your largest revenue-generators in terms of current business, but generally they also offer your best opportunity for new business and future growth**. These are your priority #1 customers.

If you've done your homework on your customers, you'll know from the outset when you take on a new customer whether they are essentially a 'one-trick pony' – a company with one product and hence only one piece of business for you – or one with a developing product range. Obviously the latter are likely to be your best prospects, potentially offering long-term strategic growth for your company, rather than the one-trick ponies.

There are other benefits to focusing on your current satisfied customers for future new business, such as cost-of-sales. **New untested, untried and unknown customers cost a lot in terms of sales expenditures to capture as a customer**. You already have a relationship with current satisfied customers, you know the people at that company, and they know you. If they really know your company and understand the value you provide, they will send you only those RFQs that they know suit your capabilities and maximize your expertise and value.

The next ring or section around the bullseye center section contains those customers that have the potential to offer you new business, but you've not quite got there with them. These are your priority #2 customers. Maybe you do one or two jobs for them and they have the potential for giving you more business, but you've not captured any more work from them. It may be that you don't know just how much future work they have to offer. Are they going to be entering new markets? Are they in the process of developing new products? This is where the cost-of-sales is a bit higher. The sales person must take the time to see where the opportunities lie with this company, and whether that means any substantial new business or just a dead end.

The third or outer ring of the target holds your priority #3 companies – those that you've identified from your preliminary research as offering good potential but you've not had any RFQs from them yet, or maybe you've had some RFQs from them, quoted the work but not yet won a contract. These require an even higher cost-of-sales, and a lot of time from the sales staff to see if you are just spinning your wheels with these companies, or to dig in and find out exactly what the problem is. Why do they keep quoting you but not giving you the work? Why won't they even send an RFQ? Are they just quoting you to keep their current molder 'honest?' Do you fit with their supply chain strategy? Those are all questions you should try to get answered. Only then do you know whether they justify any further sales effort on your part.

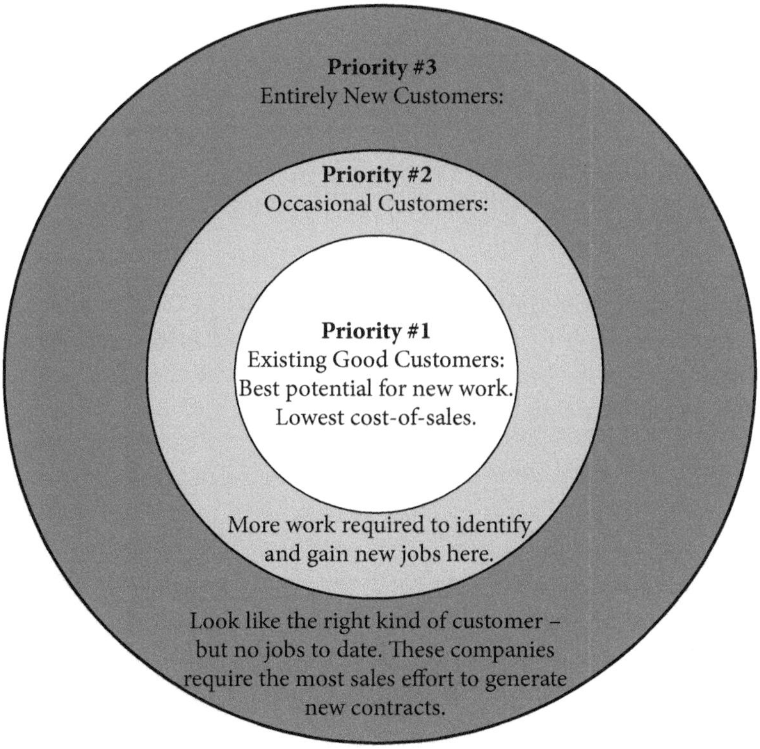

Figure 4.1

Target your customers – focus on priority #1

4.4.1 Prioritise your sales efforts

> ☞ **Keep in touch with your good customers – their plans can be your opportunities.**

Devoting the right amount of effort to the right customers is something many companies fail to take into consideration, frequently because once a prospect become a customer, the emphasis on generating new business disappears from the equation. **The focus should always be on those of your current satisfied customers that represent the largest percentage of your annual sales.** But even with satisfied customers, your sales staff can't rest on their laurels. They must be continually working on the relationship, promoting your value and keeping up with what the customer is doing in terms of future business.

Your customers will often change places from time-to-time, with some of the largest customers falling back as their business drops off, and some of the ones in the second ring becoming more important as their business grows. Potential customers will move into the 'customer' rings as you get their business. That's why it takes a lot of attention from the sales staff. Some molders have account managers to handle that center bullseye section, so that their focus in completely on those customers that represent the biggest chunk of your business. For the second and outer rings, you might want to assign them to specific outside sales people - either direct or reps - supported by an inside sales person to track RFQs, do follow-up phone calls, and generally help build up relationships.

Typically though, priority #1 customers get most of the attention on the production floor but little from anyone else, even the sales staff. A dedicated account manager for the best customers can keep up with their future plans, and make sure you are ready to respond to them.

Opportunities abound for communicating with customers and the costs need not be high. The prime routes include your website, e-mail newsletters, trade shows and more. As important as it is to spend time seeking out new customers from your potential customer data base, staying in contact with your current customers on a regular basis is equally important - if not more so. One prime chance for communication with current customers lies with your customer service staff. Usually this is the person who receives calls regarding purchase orders for parts, scheduling changes, inquiries about shipments and so forth. These calls can be excellent opportunities to ask about new requirements in the near future, increases or decreases in orders, and other information that can prove very valuable to the sales staff.

4.4.2 When one customer equals too much business

> ☞ **A customer represents too much business if the loss of that customer could put you out of business.**

We talked about the 80:20 rule (i.e. 80% of your business generally comes from 20% of your customers), but what happens when one or two customers become 80% of your business? That can be a problem. So how much business from one customer is too much business? Some industry sources say that a customer represents too much business if the loss of that customer could put you out of business. Others say you don't want one customer to be responsible for more than 20 to 25 percent of your sales volume. More than that, the logic goes, and the customer owns you.

Because every company's situation is different, only you can determine just how much work from a single customer is excessive for your company – hard and fast rules don't exist. However, some guidelines should be established – **you don't want to find yourself devastated financially when a program ends or a customer pulls the work for some reason**. Being a custom molder means that sensible diversification must be the name of the game. This is not the same thing as being all things to all people. While you can have a niche or specialize in specific markets or types of parts, you should keep in mind that a diverse customer base can reduce the risk of harm to your business if you lose work from one or two of your customers.

Equally, diversification in this instance doesn't necessarily mean doing something entirely new. Maybe you can stick with the products and technology you know, but spread your net over a wider area. Alternatively, you may need to take a good look at your current customers, and work out which other market sectors might value similar qualities (or perhaps working practices or ancillary processes) in a molder.

In today's economic climate, more OEMs are becoming cautious about the security of their suppliers. They want suppliers that have strong financials, a good banker, and the ability to manage cash flow. Too often in the past few years, OEMs have found themselves in the position of having to rush into a molded parts supplier and haul out their molds because the molder is closing the doors due to a bad financial situation. To avoid this awkward situation, and the potential delays this can cause to their own business and customers, some are now asking for an 'open book' policy before letting work. If they see you have more than 20 or 25 percent of your business with one customer, that's a red flag to many of them.

A number of years ago, a major business equipment OEM had a policy that they never wanted to be more than 25 percent of their molder's business, not because it put the molder in a tenuous situation but because it put pressure on the OEM to be nearly the sole supporter of the custom molder. That was a situation they wanted to avoid.

4.4.3 New customers – same markets

Examining those core customers with whom you have the most successful relationship can tell you a lot about where you should be looking for new business from new customers. One good rule is '**seek greater market share by locating new customers in the same markets as your current satisfied customers.**' So consider what makes you successful with your current core customer base? What capabilities do you have that other customers in that market might find valuable?

Although most industry segments have similar quality requirements for good manufacturing practices (GMPs), each segment has its own set of policies, procedures, and practices; its own set of individual requirements. Did you ever wonder why so many molders who serve the automotive do so almost exclusively? Or why many molders who do not serve that niche say they wouldn't want to? Essentially, the molders that serve the automotive market have had to become very good at meeting automotive requirements. They have the financial wherewithal to deal with the automotive industry's sometimes punitive payment terms, and the cash flow to withstand the ups and downs of that industry. They have become experts in dealing with that industry and know what red tape and paperwork are involved, as well as the quality criteria.

Those who serve the military, aerospace and defense industries are much the same. They have the military manufacturing certifications required and the expertise in difficult-to-mold engineering thermoplastic resins that many military parts require. The same holds true for the medical market, with the FDA certifications that are required for certain components used in the medical/surgical market.

Selling to potential customers with requirements similar to those of your current customers is easier and less costly than trying to penetrate unfamiliar markets. So if you're already good at molding clear crystal styrene for the medical diagnostics market, then perhaps you'll also excel at molding syringes. But you might also be well-placed to go outside the medical market to the cosmetic packaging industry where small, high-quality, transparent containers are also required.

4.4.4 New business from mold transfers

According to Jeff Mengel of Plante & Moran PLLC (**www.plantemoran.com**), mold moves have been big business recently, especially during the recession years of 2008-2009, when the number of bankruptcies among molders and tier one automotive suppliers increased. Mengel predicted that in the US in 2009, approximately $11 billion in sales and 240,000 tools would find new homes and that the trend would continue throughout 2010, projecting that more than 360,000 tools would eventually transfer.[2] With the rate of molding companies closing at an all time high in the US, many OEMs found themselves having to get their molds out of a molder's facility in a hurry. That means finding another molded parts supplier that can take on a lot of molds and get them up and running fast.

Transferring molds from one vendor to another or from one country to another isn't a easy job. However, there are some ways to make that easier, especially if your company is on the receiving end of a major transfer. Companies such as Tooling Transfers (**www.toolingtransfers.com**) specialize in the management of complex vendor-to-vendor transfers of molds and related tooling. The company promises to make mold transfers less painful, no matter where in the world you need to take them. Owner and President Paul Mulville already had extensive experience in transferring large tooling packages when he conceived of his business in 2008. He recognized that most companies did not have experience of frequent tool transfers, and opened Tooling Transfers with the express purpose of serving OEMs, mid-sized and start-up companies that have limited technical resources.

Mold moves are not usually easy and there are many critical factors that need to be considered to make the move smooth and to cause minimum disruption to the supply chain. One thing that can make mold moves awkward is a problematic relationship between the OEM and the original molder. In many instances, tooling transfers don't happen under the best of circumstances. Besides the need to change molders for simple reasons of price or geographical convenience, transfers may happen because of quality issues, disputes over payment, mergers, acquisitions and bankruptcies – all potentially challenging circumstances.

An experienced intermediary can smooth the process by applying the following types of practices recommended by Mulville:

- Understand the capacity at the new molder. It's not just understanding capacity but also the range of tooling that you'll be putting in and what implications that will have on capacity. Will there be capacity constraints? You need to identify the capacity and the immediate resources available at the new molder.

2 Mengel J.F., Leveraging the Future: How Processors Are Responding to the Recession, **2009**, http://plasticsnews.com/forum2009/presentations/JMengel.pdf.

- Build a team at the receiving molder's end to respond to the needs of the OEM – to have processing engineering, quality control, and other critical personal in place so that the molds can be received, processed, and readied for production in a timely manner.

- Understand what the deliverables are to the customer and provide a full reconciliation of the tooling. *"Understand that what you're moving and the ability of the tooling to make the required part don't always align,"* Mulville says. *"Many times we need to make corrections to the tool to match the drawing to get the part required."*

Taking on a large number of molds from another molder, while a good way to get new business, can be costly and requires care. Machine capacity needs to be considered, as well as the cost of getting the molds opened, cleaned and brought up to good running condition. A company like Tooling Transfers can be a big help.

Case history: New Berlin Plastics

Some molders cringe at the thought of taking in molds from another molder, given all the unknowns that come with mold moves, not to mention the typically adverse circumstances under which many molds are moved. New Berlin Plastics is an exception (**www.nbplastics.com**). The company has a long history in taking in transfer tooling and doing it successfully. In fact, some of its largest customers were first established with the company through a major mold move. *"We've never been reluctant to take in tooling in this way, and in fact we embrace it,"* says Thomas Frank, director of sales and marketing for New Berlin Plastics. *"We take ownership of the process, and of the mold, working to make the transition smooth for the end customer, which is why we created our Smooth Move Mold Kit."*

The Smooth Move Mold Kit provides a detailed documented structure, designed to establish an orderly process for transferring tooling and get conforming parts to the customer in the shortest amount of time possible, explains Frank. *"Back in 2006, a local molder went out of business and we stepped in to help their customers out. Through that process we transitioned over 100 molds to our facility, and had them up and running in a matter of weeks."*

Prior to the molds hitting the dock at New Berlin Plastics, the company assigns a project leader to the team of design, tooling, quality and processing engineers, to support the move from receipt of tooling through production approval. Once the tools are in-house, the team performs a thorough evaluation and takes photos of the molds, and provides a report to the customer. Next, the team has a face-to-face mold review with the customer. *"Then we sample the mold to determine if the mold makes conforming parts,"* says Frank. *"We want to use scientific injection molding to come up with a solid process. Based on the trial, we take sample parts and compare those to the sample parts from the previous molder. We have access to the capability of scanning the parts to see where the discrepancies might be."*

All these steps are set out in the Smooth Mold Move Kit, which is designed to formalize communications and eliminate misunderstandings. *"We're trying to set their expectations to be reasonable from the beginning, to help them understand what they'll be going through to get their mold up and running. That's the goal. We're trying to give them the tools to make this a smooth transition."*

Frank adds that it is not uncommon for customers to call and ask if they can get a mold into production in two days. *"This includes quality checks and performance testing to approve the product and meet our customer's tight delivery window,"* he says. *"We pride ourselves on being flexible, fast, and customer focused."*[3]

ToolingDocs (**www.toolingdocs.com**), a subsidiary of Progressive Components, also specializes in intermediary services for molds that are being transferred from one molder to another. Its MoldTrax program, developed by Steve Johnson, provides a service that includes taking in molds, opening them up, evaluating them, performing any required cleaning or maintenance, and then testing the molds to see if they make conforming parts.

Using third party services like these prevents companies from playing the 'blame game' – with the new molder blaming the former molder for mold damage, poor mold conditions or other problems that prevent the mold from running conforming parts. The intermediary informs both the OEM and the new molder about what they've got, what needs to be done to get conforming parts, and what it will cost in both time and money.[4]

It's certainly tempting to look at a major tooling move as a boon for the receiving molder, but it's critical to calculate the cost of getting these molds, and the time and effort needed to ensure that they can be up and running in a reasonable time period.

---

3 Goldsberry C., Need to move a mold? How one molder makes it smooth, January 11, **2010**, www.plasticstoday.com.

4 Goldsberry C., Moldmakers play important role in successful tool transfers, September 14, **2009**, www.plasticstoday.com.

4.5 Internal communication is essential

It often seems that sales people are first to be blamed when things aren't going well and business is down. Sales people – particularly outside sales people – are perceived as never available because their job demands that they be outside the plant so often. But these people are often thought of as off having a good time, enjoying nice lunches, playing golf (we've already discussed the potential benefits of this!), or sometimes generally goofing off. What every sales organization needs to do is provide upper management and the people in the plant who have a vested interest in sales activities, with a weekly report on the sales activity.

Upper management needs to know which companies are visited, who the sales person is seeing at that company, what is being said, what opportunities are currently available and/or what opportunities might be coming in the near future. In other words, management needs a detailed accounting of all sales activities. Now this may rub against the grain of the sales people, because they don't like being bogged down in paperwork, but it is absolutely necessary to have this type of communication if the sales person is to be successful.

It is also necessary to communicate this information with the people in the plant: such as molding operations managers, quality managers and engineering managers (who may or may not be actively involved in sales). This is particularly true when things start getting slow. The people on the plant floor have absolutely no control over the work that is coming in, and if work is sparse they begin to get worried. To know that there is a sales team working to promote the company and its capabilities, and beating the bushes for new work provides them with the knowledge that the company cares, that new work will come soon, and that their jobs are safe.

One main reason why this type of communication is important is that the sales cycle for custom injection molding business is quite long. It takes many months – sometime even a year – to develop a relationship to the point of getting a molding job. Many OEMs buy molds only periodically, and the mold buying cycle might have just passed, so it may be six months before another opportunity comes up. If so, this is important to communicate with upper management.

Also, others within the company can help the sales people solve the problems they encounter with selling, and evaluate the situation. When a sales person fails to secure a job the following questions might be considered:

Why did you lose the job? Typically the customer will say "*the price was too high*". If this really seems to be the case then maybe management and engineering needs to sit down and talk about the situation.

"Are we really competitive? Is this really where the market is right now?"

"If so, then what can we do to get the price down? Take the man-hours out? Reduce machining time?"

Talking about the RFQs and what it takes to win a job is everyone's responsibility, and everyone needs to work with sales to make the jobs happen. **The sales team don't have control over the OEMs and when or why or how they let jobs to their suppliers. Similarly, they don't have control over the price of the mold or molded parts, or how long it takes to build a mold and get it production ready. The sales person only has control over the time and effort he or she puts into creating relationships and opportunities to build new business.** That means that it's everyone's job at the plant to provide the best solutions, the best resources and the best brainstorming with the sales department so that everyone can be successful.

Chapter 5

Business planning for success

The future for any business is always uncertain, and during the first decade of the 21st century we've seen what were believed to be solid, growing companies fall into bankruptcy, and some even go out of business. Economic issues plaguing both North America and Europe have impacted even the smallest molding companies, and given cause for many owners to rethink their business strategy – or at least to implement a business strategy.

If you have not yet implemented a marketing and sales strategy, then these should be high on your action list – they truly are the road map to getting where you want to go with your business. **Along with the marketing and sales strategy, you absolutely need a business strategy or plan to ensure you keep doing what you set out to do**. There are no 'accidental molders.' You can't just buy presses, hang out a sign and hope the world will beat a path to your doorstep.

There is a world of competitors out there, and the competitiveness issues that remain a thorn for molders and moldmakers in Europe and North America aren't likely to go away any time soon. However, times are changing, with rising wages, striking workers, and other 'Western' ideas beginning to take root in lower cost countries, and many OEMs are at least reconsidering their decisions to move their manufacturing offshore purely for the benefits of cheaper labor.

5.1 Being successful in downtimes

Learning to take advantage of business cycles can bring real opportunities for molding company owners. Too often, when the economy or the industry go into a down-turn, people panic and freeze. After all, when times are good it's easy to succeed. But when things begin going sour, it's a different story. Everyone wants to 'turtle' – pull into their shells and wait it out. Yet, an industry downturn can bring new ways to grow and direct your future.

For example, companies going out of business can increase the availability of good, used equipment, and might provide the opportunity to take on some secondary operations without the high cost of investing in new equipment for welding, hot stamping, etc. Buildings repossessed by banks and lending institutions can be purchased for less than market value, offering other molding companies the opportunity to expand their operations.

Some molders who are planning ahead (as everyone should) use the boom times to strategize, and then when the downturn hits, they can make other molders' misfortune their own opportunity. Companies who have strong business planning will use the good times to pay off debt and save up for the down times in the business cycle, something they know will eventually come along.

It is also wise in the good times to keep looking for new customers, because when everyone is busy they're not out looking for more work. A smart company goes after those valuable customers they know may take a long time to get. Besides, when you're really busy, you don't have to have immediate business, and you can pick and choose those customers you really want, and by-pass those that aren't a good fit.

The boom times also allow you to cull those customers that aren't profitable to make room for new customers that are. So don't 'turtle' during the down cycles or rest on your laurels in the good times! There's a lot you can be doing. But these things don't happen on a whim. You can't do this flying by the seat of your pants – you've got to plan, plan, plan!

Ten Ways to Take Advantage of Tough Times by Patrick McNally, MBA, CPA and partner at Blackman Kallick LLP (www.blackmankallick.com)

Patrick McNally works with many small-to-mid-sized businesses, including many molding companies. He says that any company can survive when times are good because "*lots of business can hide a multitude of sins.*" However, bad times will weed out the weak, the poor performers, and provide opportunities for the strong to get stronger. He made the following suggestions in an article back in 2002,[1] but his advice is just as relevant in any recessionary times:

1. **Buy Equipment**. We talk about this opportunity in a couple of sections in this book, but it's true. You can get some good equipment without making a huge investment.
2. **Upgrade staff**. When times aren't so good in manufacturing, there are often many good people unemployed.
3. **Acquire a troubled competitor**. McNally says that you might be able to get a 'real deal' by acquiring a troubled competitor – a company that has something to offer that will help you expand your business, such as customers in a different market, different capabilities or different sized presses. "*One benefit of buying a going concern is that you gain assets such as trademarks, know-how, and intellectual property, and for the price of the bank debt, you can acquire inventory and equipment – and essentially get a list of customers for free*".

1 McNally P., **2002**, www.blackmankallick.com/articles/2002/11/10-ways-to-take-advantage-of-troubled-times/.

4. **Acquire a troubled vendor**. McNally says that depending on your industry, it may offer a competitive edge to secure a source of supply, and bring that capability in house. For a molding company, acquiring a vendor of silk screening or pad printing, for example, might help add value to your own offer.

5. **Pursue troubled competitors' customers**. "*Find out who your troubled competitors are and focus on their customers.*" However, there is a caveat: "*It may have been those customers who got your competitors into trouble in the first place, there's nothing to be gained by pursuing their troubled customers.*"

6. **Fire customers**. We've also talked about this in a previous chapter. McNally says this is a good thing to do, "*especially if your cash is tight, you need to take a hard look at your customers and make sure you're making money on each one of them. Consider raising prices and reducing terms on your marginal customers.*"

7. **Negotiate long-term price concessions from suppliers**. Renegotiating long-term or contract pricing has advantages, but just be sure you're not fixing prices at a high point. Pick your suppliers carefully when re-negotiating.

8. **Take advantage of cheaper advertising**. When times are bad, many of your competitors will pull back on their advertising (an example of turtling), when in reality they should be doing more. However, the up side of fewer ads is that pricing deals might be more readily available. "*When people consistently see your name – and not those of your competitors – you will be perceived as a winning company.*"

9. **Renegotiate long-term leases**. During downturns, especially those that affect real estate, such as the most recent 2008-2009 downturn, you might be able to successfully renegotiate your lease for a longer term. "*You'll help yourself by getting a lower rate than you would if you wait until the property market runs around.*"

10. **Evaluate your strengths and weaknesses**. We've talked about this in our strategic marketing planning, and while any time is a good time to do a SWOT evaluation, one of the optimum times is during a downturn. "*It's the perfect time to evaluate what your company does well and focus on those strengths – and pinpoint your weaknesses and fix them. Tough times tend to focus people's energies; employees get serious about making improvements when they're afraid of losing their jobs.*"

5.2 Business planning

What do you want your business to look like five years from now and where do you want to be? The vast majority of custom injection molding businesses are privately held companies with a single owner or a partnership, so it's important to put in place a business plan that provides some direction for the company, its owner(s) and stakeholders (employees). Too often molding company owners are so busy growing the business and taking care of the day-to-day challenges in business operations that they rarely think about the future. The business grows, things and people within the business change, yet no plans are ever laid to address the growth and the changes, and their impact on the long-term outlook of the business.

> ☞ **Too often molding company owners are so busy growing the business and taking care of the day-to-day challenges in business operations that they rarely think about the future.**

5.2.1 What is a business plan for?

Business planning comes in handy for many reasons. The main reason is clearly to ensure that you have a workable and costed strategy for the future of your business, however banks will often ask to see your business plan if you ask them to extend your line of credit or finance machinery and equipment. It shows them that you're actually thinking about your business, not just riding the wind whichever way it blows. It shows that you have goals with respect to your business, and a strategy for achieving them. You may also want to share your business plan with current employees, new recruits joining your company in senior positions, or potential investors in your firm.

One good way of writing your business plan is to ensure that it provides all of the following things:

1. **A compelling story** showing where your company has come from – who founded it; where and when it was founded; why it was founded; how did the business develop, evolve and grow over the years; and why has your business been successful?

2. **A clear picture** of your company as it is now – don't ramble – focus on what you own and what you do.

3. **A clear vision** of where you want your company to go in the future, in the form of defined, specific business objectives and goals, together with ways

of measuring and monitoring whether you are on track to achieve them. Business objectives should be more than just a 'wish list'; you need to include realistic dates and deadlines.

4. **A written plan** that forces logic and gate-keeping into the business – It's one thing to talk about your goals and objectives, but it's really important to write out your business plan. Looking at it on paper gives you a new perspective and also forces you to follow it – providing clarity to your vision. Writing down your objectives and timescales can also help you to calculate the manpower and investment required to make the changes you envisage, and really evaluate their feasibility and potential return.

5. **A living document** – the process of writing a business plan is valuable in its own right, but just because you've got it written out doesn't mean you file it away in the bottom of your filing cabinet never to look at it again. A business plan is a living document that must be taken out at least once a year, reviewed by management and updated as needed. Businesses change over time, people change, your vision will change as you move into the future, so don't allow your business plan become stagnant and obsolete.

5.2.2 Writing your business plan

Remembering that you may share your business plan with trusted colleagues inside or outside the company can help you to decide how much detail to include. For example you might be considering expansion through acquisition of another company. Your business plan could include details of the markets or regions you hope to target through such an acquisition, or the technical expertise you hope to gain. However, you would not include names of possible target companies.

You should also take into account that you may not just be writing your business plan for industry insiders. A little general background information can help someone like your bank manager understand where custom molding fits within manufacturing industry.

Writing a business plan is the responsibility of management, who must make the necessary financial decisions to convert plans into actions. However, there is often valuable knowledge elsewhere in a company, and you should find a way of feeding this knowledge into the planning process. Shop floor staff may know that a key piece of equipment is due for replacement, and your salesman may know that a key customer is making changes. Purchasing may be aware of impending materials price rises. Any of these could affect your ability to achieve short-term goals, and this type of information should be routinely gathered and monitored. Several documents have been described elsewhere in this book which should feed valuable information into your business plan. Responsibility for compiling supporting information may be delegated.

There are about seven components to a business plan. You might have one or two more or less, but that is the average number of sections that comprise a good business plan. Those are as follows:

1. **The Executive Summary** – This should be written last, but is the first section. Why put the Executive Summary first? This is just what it says – it's a summary of the Business Plan and typically runs two pages in length, and provides a snapshot of your business and your goals. The executive summary allows the person reviewing your plan to quickly know right up front the key features of your company: who you are, what you do and why. It should show a positive approach that conveys your knowledge of your business, and your confidence in your business and yourself.
2. **A Description of the Business and Your Vision** – This can be developed from the **SWOT evaluation** that you performed for your **Marketing & Sales Strategy**. Bearing in mind your audience, i.e. bankers and others who may not know what molding is, your description of your business should include an explanation of what injection molding is, and how being a custom injection molder brings opportunities to participate in the manufacturing of many types of components in many different industries – note how ubiquitous the plastics industry is from cars, trucks, and airplanes to medical devices and equipment, toys, sporting goods, agriculture, recreational products, telecommunications, business equipment, computers and more. Your vision for your business however, should be focused, and again you may cull this focus from the work you did on your strategic marketing and sales plan. Take those sections from your SWOT that determine what you do best and who you do it best for and create a vision for your business based on this information.

   Your vision should contain **short and long term objectives** – perhaps how much you would like to increase your turnover in one year and in the next three or five. The detail you then add in the following sections will then help you to determine and explain precisely how you will achieve this growth. It is difficult to plan realistically for any longer than five years as a custom molder, with so many external factors and variables: materials pricing changes constantly, markets change, OEMs get bought and sold and switch suppliers in a heartbeat. The important thing is to revisit these goals annually, and reset them if necessary in the light of all these factors.

3. **Definition of the Market** – Here you can go into the markets you serve, by geography and/or business sector, and what makes those ideal markets for you. How are these markets doing in today's economy? What opportunities do they offer your business to grow and be profitable? What will they expect from you in the future? Provide some research information on the markets you serve (available from many different industry research firms such as AMI, The Freedonia Group, Frost & Sullivan, and also trade associations and similar groups). You should also be able to show with facts and figures that these markets are thriving, and that you have the services, capabilities, and technologies that they will require moving forward.

   This exercise may show that you need to be looking for new markets, and again you need some clear plans and goals. These could include short term plans to join exploratory trade missions or attend trade shows in new sectors. In the longer term you might aim to make 30% of your turnover from a more lucrative industry sector within perhaps five or ten years.

4. **Description of Services & Technologies** – This is where you can talk about the various services you perform – the primary one being injection molding – but you may also build molds, or add value downstream through assembly or decoration for example. You might want to describe what injection molding is – nothing real long or technically detailed but a description for laypeople who perhaps can relate to your description by looking at their computer keyboard or their mobile phone. If you are also a moldmaker, you might also explain how manufacturing of the injection molds is a critical component of making the molded parts. Next, explain the equipment that you have to perform injection molding, and the technology required to injection mold parts in a manufacturing world that demands greater efficiencies, higher quality, faster delivery times, and a lot of value-added services (and the equipment that is required to perform these).

   A detailed **inventory** of the age, capacity and capability of all the equipment in your plant is beyond the scope of a business plan. However this is vital information when planning for investment and growth. Do you have the capacity to increase throughput as you have defined above? Will new markets or products require investment in greater automation, ancillary equipment or even new presses? An inventory of all your plant, together with records of repairs and maintenance, may also flag up instances where you should be planning to replace older or worn-out equipment. Your business plan should include realistic figures for investment in plant and equipment.

5. **Organization, Management and Personnel** – Again you can start by defining what you have, but this time looking at the people you employ, the jobs they do, and the skills they have. Draw up an **organisation chart** showing your management structure, and their responsibilities for

production, quality, sales, marketing, etc. In a small firm this may be a very simple diagram. This may also be a good time to ensure that you have allocated responsibility for driving forward any new initiatives described elsewhere in the business plan. Then you may also want to map out what you have in terms of production, administration and support staff, and the key skills they contribute to the success of the business.

If you carry out a **skills audit** as described in Chapter 6, then you will probably identify some gaps to be filled by recruitment or training. Similarly, any plans to develop new products or markets may also bring a need for new skills (but new working practices can also free up personnel by reducing their workload). Your business plan should reflect all these potential changes, and include realistic budgets and timescales for training and recruitment.

6. **Marketing & Sales Strategy** – Marketing and sales are just as important to the future of your company as the services and technologies you deliver, so this should be an equally important part of your business plan. Here's where you'll be happy you've completed your Marketing & Sales Strategy. You don't have to present the entire strategy, but you can provide an executive summary, and focus particularly on those areas requiring the greatest innovation or investment of time and resources. Anyone reading your business plan will know that you've actually taken the time to think carefully about your business, what it is you do best, who you do it best for, what are the most profitable opportunities for new business growth, and how you're going to make money and grow the business. Now this is part of your business plan you should also make sure that you have included specific targets and deadlines, and calculated the costs so you can allocate realistic budgets.
7. **Financial Management** – This is always a key element because you need to provide a snapshot of your finances, your inventory turns, how you manage your accounts receivable and payable. Do you manage your company's finances well? Do you have many outstanding receivables? Are your payables up to date? Are you on COD (cash on delivery) with any of your suppliers? These are all critical pieces of information to your business plan. If there's something here that doesn't look very good, then you must plan to correct it so that your business looks better financially.

There are many software programs that can help you develop a business plan by walking you through the process step-by-step. Also check online for free guidelines to writing a business plan. While they won't apply specifically to injection molding, you can tweak the guidelines to make them applicable to your business. **Remember, it's not about the number of pages in your business plan – you don't have to write 'War and Peace' – it's about content**!

5.3 Expanding a business through acquisition

If your future calls for an expansion, you also might consider buying another company. This can bring you new plant and equipment, new customers and possibly entry into a new market sector or geographical region – without having to start from scratch. Your business plan can help you to decide where to start looking. **Do you want to offer 'more of the same' in a growing market? Can you add value to your overall business offering by expanding the range of machine sizes you can offer**, maybe looking for an acquisition who specialize in very large, small or highly complex moldings for example? Alternatively you may want to **expand your offering into downstream added-value operations** such as decorating, or even pick up a proprietary product line which is complementary to your business. The better you understand your existing business and goals, the better you will be placed to evaluate a potential acquisition.

Companies come onto the market for many reasons, and this is particularly true in recessionary times, when you may even contemplate acquiring a former competitor that has gone bankrupt. The current economic recession has provided many such opportunities for molders to expand their business particularly if they are looking to grow into a new geographic region, or diversify into a new market. There are molding companies that are on the edge of going out of business, or those who are in bankruptcy will little hope of getting out, and need to sell quickly.

Buying a company in bankruptcy can be advantageous, although it has its dangers. First, it means that you'll pay less than for a company that is a thriving entity. Typically, you'll pay a premium for a successful company, so taking a hard look at a bankrupt company might be a good thing. Patrick McNally from Blackman Kallick has said that often companies overlook a bankrupt firm because they don't understand the opportunities that exist in that situation. However acquiring one of these companies is a very inexpensive way to pick up additional sales or to acquire a whole new business. You are buying a company virtually for its liquidation value. Also there are various aspects to buying a bankrupt business. For example, if you don't want the equipment, then he suggests you should try buying just the troubled company's accounts receivables or its customer list. However, he points out that the purchase is generally done quickly when you're buying a troubled company, and you don't have the luxury of taking 60 days to conduct due diligence.

Buying a successful business – another molding company in a market in which you'd like to serve, or a supplier such as a company specializing in secondary operations or one that specializes in product design and development – is also a way to expand your capabilities and extend your reach into areas many molders don't typically have in-house. Often you can find companies listed for sale with companies that specialize in mergers and acquisitions. You might also come across a company for sale just by networking in the industry and letting your colleagues know that you're looking for a company that will help you extend your reach. Successful small businesses can come onto the market when the owners are contemplating retirement, or just a change of direction.

As with any major business decision, you need to enlist the help of a reputable business broker, your accountant, and others who know the commercial side of business acquisitions. A good team of knowledgeable people can provide you with the due diligence required to help you make the right decision so that you're not buying a business with hidden problems.

Case history: Dickten Masch Plastics

There can also be opportunities within your customer base to purchase a business. One of Dickten Masch Plastics' (DMP, **www.dicktenplastics.com**) customers for more than 25 years, Charter Automotive, decided to stop making a fluid level indicator (a dipstick) that measures various fluids in vehicle engines, so DMP explored making the product itself. DMP's management spoke with Charter about approaching their customers, developed a business case and brought its combined teams together getting everyone involved in their area of expertise, resulting in a quick acquisition. In addition to securing the new customers and business, DMP purchased the equipment and took on Charter employees. Today Dickten Masch Plastics business is about one third automotive (tier one and tier two). They have added TS 16949 certification in January 2011 and are the dominate producer of dipsticks in North America. The business was secured in the economic downturn in 2009 when the automotive run rate was 8 million units and today it is trending to 15 million units.

Chapter 6

Employee training, hiring and firing

Ask almost any molding company owner and they will tell you that their most important asset is their workforce. It is the human talent, experience and expertise that make an injection molding or moldmaking company run and run well. By the same token, poorly trained and inexperienced employees can cause a lot of problems including poor quality parts, damaged equipment and lack of motivation.

And if employees are indeed a company's greatest asset, then there must be continued investment in your employees to keep them motivated, up-to-date with training and equipment operation, and to help them to feel 'ownership' of their jobs and that they are a part of the company's success.

In the United States alone, there are approximately 1.1 million people employed in the plastics industry, responsible for $379 billion in annual shipments, according to the Society of the Plastics Industry. In Europe, there are more than 1.6 million people generating an annual turnover of €280 billion, according to EuPC, the European trade association for plastics converters. Most are employed in small and medium sized businesses, but together these represent a significant contribution to their countries' economic performance, demonstrating the importance of maintaining competitiveness and efficiency from the shop floor upwards.

6.1 Training for success

Employers need to provide formal training programs to their employees as a means of keeping their companies competitive. Douglas K. Woods, President of the Association for Manufacturing Technology (**www.amtonline.org**), suggested to a group of the American Mold Builders Association in September 2010, that manufacturing companies need to provide a minimum of 40 hours of training annually for their employees to keep up with new manufacturing technology.

In most cases, there is a direct correlation between the level of employee training and a processor's productivity and efficiency, and the quality of the products the company produces. And most of the critical issues in training are concerned with new technology in machinery and materials, according to Mike Dalton, director of quality for contract molder MRPC Corp.

(**www.mrpcorp.com**).[1] Changes in materials including new compounds, new molding parameters for new materials, and changes in machine technology including robotics and other automation, mean that employees should continually update their knowledge base to keep up.

"We constantly need our employees to perform at higher levels of technology," Dalton said. *"Basically, we're only as good as our worst operator, so we carefully screen employees prior to hiring and then screen them during a 60-day training and orientation period."*

MRPC has 90 employees, and when the company makes new hires, a new employee's first few days are spent in the classroom, training on topics such as safety, plant operations, and what it means for a company to be ISO certified. Once the new employees get on the production floor, depending on which department they are assigned to, the assigned trainer will give the person more and more responsibility over time as their training progresses.

Another element of the training program is insight into the use of the parts being molded. **A part is a part is a part** to an operator who might not have a clue as to what the component is ultimately being used for. An understanding of the function of the part, which might not look like anything familiar but eventually goes into a product that everyone uses, helps the operators and quality people see the importance of their job. **A small component could be destined for a safety-critical industrial application, a high performance vehicle, or even a life-saving medical device.** *"Often the only thing standing between a conforming part and a nonconforming part is the press operator,"* said Dalton.

Many molding companies use a 'cross-training' program to provide their employees with additional knowledge, and to give them opportunities in various departments, even if it is just to fill in if another employee in that department calls in sick. At MRPC, each operator trains in quality as well as processing. *"We use what we call a Production Tracker that goes with the job,"* said Dalton. *"It's a requirement that our operators read this, then learn to translate what's on the paper to the process, which is key. Because our operators are trained in the basics of processing, they use the Production Tracker, which tells them the parameters of the press, criteria of the product, and other information they need to run the job successfully."*

Andy Routsis, president of A. Routsis Associates (**www.andyroutsis.com**), a company that designs and administers training programs for plastics processors,

---

1 Goldsberry C., Want good products and processes? Train good employees, September 15, **2009**, www.plasticstoday.com.

said that training programs have taken on a greater level of importance during the past few years. "*Molders realize that if they don't train their employees, they'll have bigger problems than they already have,*" he said.

> ☞ ***"The only thing that sets you apart is your knowledge and utilization of that knowledge. Everything else – the machinery, software, and equipment – your competitors can buy."***

Some of the reasons for this, Routsis speculates, is that due to the retirement of many people in the molding and moldmaking industry, a lot of knowledge base has been lost, resulting in molders demanding more highly educated and skilled employees. Add to that the number of people who've been laid off, and you've got an even greater knowledge gap to contend with. Lost knowledge at the OEMs as they've downsized means that more and more OEMs are relying on their mold and parts suppliers to fill in those gaps, forcing many molders to hire experienced personnel. Fortunately, some of those being laid off from the OEMs are finding a nice niche with the molding and moldmaking industries.

While many molding companies look for the tangible benefits of skilled workers such as **reduced scrap rate or reduced downtime on machinery** because those can be measured, there are the intangible benefits of experience and training. Routsis said that these include **reduced tooling and equipment damage**. The number one intangible is communication, Routsis said. "*This is big, because communicating with everyone as to what the problems are, how they're being solved, and what needs to happen is critical to the process,*" he said. ***"Morale is better with training because employees understand what their path is, which means accidents and absenteeism are reduced."***

RJG Inc. (**www.rjginc.com**), a Michigan company that provides training programs to improve efficiencies and productivity in molding, has seen an increased demand for their services as a means to reduce overall cost-to-manufacture. One of RJG's trainers said that with the huge investment in equipment, it only makes sense to also invest in the people that will be running the equipment. "*The only thing that sets you apart is your knowledge and utilization of that knowledge,*" said this trainer. "*Everything else – the machinery, software, and equipment – your competitors can buy.*"

While the cost of training programs might be high, and some molders do question their return on investment, MRPC's Dalton believes that the ROI for its employees has been huge. "*The reason we spend so much money on training is we know that if employees don't do their job – don't know how to do their job – it will ultimately cost us a lot more. Scrap dollars are like free money. Reduced scrap is money in the bank.*" It also results in better parts for the customer, and after all, that is what all customers want: quality parts that meet form, fit and function requirements. A molder that can deliver good parts on-time and at competitive prices will reap the benefits of a long-term relationship with its customers.

That should be enough to get management buy-in. As with any program, management buy-in is critical to its success. Training has to be from the top down, said these training companies, or it won't be successful.

So how do you work out what training is required, and get the best value from your training budget? Appendix 6.1 contains suggestions for carrying out a skills audit and devising a training plan.

Finally, it must be acknowledged that lay-offs are a common occurrence, especially in the global recessionary times at the beginning of the 21st century. They are always a possibility and just one of many reasons why employers should provide training, and particularly why employees should also be responsible for updating their skills or getting an education. According to one HR consultant, employees should always be encouraged to keep up their skills or learn new ones to help them stay 'employable.' Some companies have implemented a strategy, not just in response to globalization, but to acquisitions and automation, to help employees recognize that they are responsible for managing their own careers, to be employable and maintain their livelihood. Many provide staff training programs for advancing in technical aspects of their jobs, and some will even help employees get degrees in areas that are beneficial to their jobs.

☞ ***"Reduced scrap is money in the bank."***

6.2 Challenging times for employees

While productivity gains in recent years have made both North America and Europe more competitive in the face of manufacturing going to low labor cost countries, much of that productivity has come from employees striving to maintain their jobs and keep the work from going abroad.

Many of the big, multinational OEMs have literally wiped many of their plants off the map in the US and Western Europe, as they've moved south and east. Much of this has come as a result of consolidation among these OEMs. For example, when Whirlpool Corp. began buying up its competitors to become a huge appliance manufacturing conglomerate, they closed many of these acquired facilities so that they could eliminate duplication of manufacturing capacity and capabilities.

Similarly Maytag Corp. had closed its large refrigerator manufacturing plant in Galesburg, Illinois, in the fall of 2002, moving the work to their Reynosa, Mexico, facility in 2003. That resulted in $111 million in lost payroll, along with 1600 jobs. The ripple effect also impacted approximately another 2000-5000 people, working for Maytag suppliers. When Maytag was acquired by Whirlpool Corp. in March of 2006, many of the Maytag plants that had survived the moves to Mexico then faced closure.

These are just high-profile examples amongst thousands of companies that have closed facilities, laid off workers, and moved plants permanently to low labor cost countries leaving behind workers feeling misplaced and fearful that manufacturing would leave Western countries for good. **While the globalization of the workforce has been good for many, including consumers, it has left many more wondering how to cope.**

This stressful situation that still exists today impacts many employees who've been forced to take lower paying jobs, or move from management positions into other positions that don't offer the pay or the prestige they were once used to. The feeling of 'community' that existed in the workplace for many is gone, as they've watched their friends and co-workers being laid off in these uncertain times. This often creates what some human resources experts call **'toxic employees'** – those who can't handle the stress and go off the deep end, sometimes even hurting themselves and/or their co-workers.

Even a company purchasing another company – with no lay-offs – can cause distress for employees. If yours is the company that was purchased, try to learn early on the intent of the acquiring company with respect to consolidations, plant closures or lay-offs, and communicate with the employees as best you can what the plans include. Will new management be brought in? What will change and how will it change? Will there be new opportunities for employees to be promoted? Or, will there be lay-offs? Not knowing can be more stressful for employees than knowing, and potentially harmful to a business – even if the news isn't good.

6.3 Challenging times for employers

Karla Dobeck, an HR consultant and owner of Human Resource Techniques Inc. (**www.hrtechniques.biz**), said that the 'losing my job' scare is pretty much a common thing in the manufacturing industry[2]. However, Dobeck said she believes in good communications between management and production

2 Goldsberry C., The work force, globalized, October 31, **2005**, www.plasticstoday.com.

employees to allay these fears, and help employees – even those who are laid off – be confident that it's not the end of the world.

"Companies should take care to communicate with their employees about issues that impact the work place," Dobeck told a group of mold manufacturing company owners. *"They can do one-on-one informational meetings, periodic state-of-the-company meetings, letting employees know how sales are doing and that the firm is looking for business in these new markets, that overhead is increasing but sales are up. Not just because of fears of globalization, this open communication makes them feel they are a part of the company, and they are less likely to invent things in their own minds."*

Dobeck also recommends being up-front about any impending lay-offs. *"If you are facing some type of lay-off, let employees know early on that this action is pending, so that people who might be thinking about leaving will go ahead and leave, and you don't have to lay off as many,"* she said.

"In the final selection of who's going, do it as a group so they won't think they're being singled out," she advised. She also recommends bringing in professional advisors to talk about issues such as benefits and pensions. Employees notice the way you've treated those who are leaving and have a better feel for the company.

Dobeck acknowledges that change is difficult, but she believes that the more employees are kept in the loop and made aware of what's going on, *"the better they will understand that the company is making a change to be more competitive, and the more readily they'll accept that change."*

Appendix 6.1: Skills audit and training plan

In the same way as you cannot embark upon a marketing program without understanding what you have to offer and to whom, so it is important before you invest in training to determine what skills you already have in the organisation, and what additional skills you most need.

A good way to start this process is to carry out a skills audit of every employee, and to collate all this information to gain an overall picture of the state of the organisation. If kept simple, this process can be repeated every year as part of an appraisal program, which can help to ensure that individuals and the company as a whole are developing and maintaining appropriate expertise. You may also choose to make this process part of a new-employee induction program, with a requirement that certain key boxes must be ticked within the first 6 or 12 months of their employment.

This survey may actually bring some pleasant surprises. Employees may have skills and qualifications gained whilst working for a previous employer, which they have never promoted or exploited since joining your company. However, a note of caution is also appropriate here. **A commitment to training will require continued expenditure. Employees may actually become demotivated and cynical if this survey is not followed through by the provision of training. Do not offer what you cannot deliver!**

The following template can be adapted to suit the specific needs of your company, for completion by every employee. It should be noted that this example is not just relevant to production staff but also to commercial, support and administrative personnel, who are just as important to the continued success of your business – if your office staff left tomorrow could anyone else order in materials or send out bills?

When adapting the template you also need to consider what skills and expertise are particularly relevant to your business, and where you want to take it in the future. For example, if you are considering moving into a high-value market sector then maybe you need greater depth in CAD or quality assurance. Alternatively if you are looking at chasing new customers overseas then maybe your sales and admin staff will require some basic language skills.

Employee name and role in the organisation:

Formal qualifications: e.g. degree, other recognised national qualification, apprenticeship:

Training received (with dates). State whether a specific qualification is required and achieved.

Training provided in-house by supervisor / peers	Training provided by supplier of machine / equipment / software	3rd party / commercial training

Job specific training

Press operation (subdivide as necessary e.g. machine setting, control systems...)			
Ancillary equipment operation (sub-divide as appropriate)			
Experience of working in other production areas			

Related skills

Materials properties and selection			
Part design / CAD			
Quality assurance			
Health and safety			
First aid			
Lifting and handling			
Energy efficiency			

General / non-technical skills

IT, accounting and general business software			
Business skills e.g. marketing, customer service			
Languages			

Any other relevant skills or training received:

Training needs identified (with timescale):

Having gathered all this information about individual employees, the next step is to collate it into one skills matrix, which will show exactly where your strengths and weakness lie across your company. You might also wish to take an empty form and decide where your priority areas are. Is it important that all machine operators have been trained in a new control system? Have you lost valuable expertise through retirement or other staff changes? You can then identify exactly where you need to apply your training budget.

Looking at feedback from individuals will also help you to decide who should receive training in a particular area. Which employees are interested in continuing professional development or have a particular aptitude? Even if your company typically relies on in-house training by peers and supervisors, you may decide that certain staff should receive third-party training on new equipment, which they can then pass on to their colleagues. Some employees may be less than enthusiastic about new technology, but have a real interest in driving energy savings.

Something else to consider is where to go for training. This can be provided in local colleges, by machinery providers, or on your own premises if you have a number of employees to be trained at once. You need to weigh up the respective costs and benefits. If you work with difficult materials or complex molds you may find it particularly helpful to bring a trainer into your own business. Some aspects of any manufacturing process certainly benefit from hands-on training, but in other areas (such as design) you may want to look at e-learning via the internet.

Once you have all this information you can draw up a training plan for the company and each individual employee – and then make sure it really happens! You need to make sure that one person is responsible for managing, organising and monitoring training, and typically this fits with other aspects of human resources (HR).If it is left to individual employees, chances are they will be so busy with day-to-day operations that training will be pushed to the back of their minds, and next time you run an audit little will have happened!

Finally – make sure you make the most of your training spend. When an employee attends a training course make sure they feed back. Design a simple evaluation form asking the following types of questions: Was it worthwhile? Did they learn what they needed to know? How can they share new skills with colleagues?

Understanding the industry

Chapter 7

Markets and profitability

7.1 Finding winning markets

The first stage of a marketing strategy will help you find the winning markets where there are potential customers who need the services that you have to offer. It is not easy to uncover these markets, which is why you have to begin your strategy by looking at your own company first. However the fragmentation of the injection molding industry is only exceeded by the fragmentation of every other industry segment out there and plastic moldings make their way into a remarkable number of them. And while there has been a large number of mergers and acquisitions throughout many market and industry segments over the past decade, research is required to sift through all those segments to find those markets, industries and companies within those that need what you have to offer.

That's why it is critical for each custom injection molding company to carefully define the focus of its business, whether that is a specific industry (automotive, medical, etc.), a specific press size (over 500 tons, 50 tons and under, etc.), or a particular specialty material such as liquid silicone rubber (LSR), thermoplastic polyolefin elastomers (TPOs), or engineering thermoplastics such as Ultem polyetherimide for example. Every custom injection molder must understand whether the dozens of opportunities that come through your e-mail each month in the form of RFQs, are truly opportunities that fit your core competency, that maximize your technological capabilities, your human experience and expertise, and take advantage of those things you do best. Recognising those will ensure a greater probability of success and your goals of profitability will be met.

But remember that nothing stands still. Existing markets can evolve or simply disappear. Once you have that clear picture of what your company currently does well, and for whom, you should keep a close watch on that business. Make sure you are aware of the trends and drivers in your markets, whether they are technologically led, commercial or perhaps environmental. Even ask yourself whether this is still a good business for your company. When you have all this information to hand, you are well placed to consider where else you might look to grow your business.

It's important to research the markets that you believe you would like to serve to see what types of products they make, the size of these products, the materials they use, and whether these are similar to your existing products. You also need to look more closely at new market areas to discover whether their supply chain is open to new suppliers and in which regions of the world.

Different industries do business in different ways, and it's also necessary to see if your company has the expertise and capabilities required before you try to enter these markets. Some sectors require much greater input from suppliers at the product design stage, or alternatively expect molders to offer a wide range of assembly or decorating technologies. Some businesses require compliance with particular manufacturing standards, such as the medical device segment, whereas some will test your cashflow to its limits with their payment terms (e.g. the automotive industry). The more you know about possible new markets, the more likely you are to see a good return on marketing expenditure.

Keep watch on your existing markets:

✓ Is the industry changing? Where are the changes coming from?

✓ Is this still a good business for you?

Look at new markets:

✓ Do they have similar requirements to your existing customers?

✓ Will you need to invest in ancillary equipment?

✓ Do they require different standards or practices?

✓ Is the industry growing and open to new suppliers?

There are many markets, many of them niche markets, but the major ones are listed below, and some interesting recent developments are highlighted. Automotive, medical device/equipment, and packaging are the largest markets for injection molders. Other markets reviewed certainly use large volumes of molded components, but several have taken almost all their manufacturing to China or other Southeast Asian countries. Others have gone to Eastern Europe or Mexico, remaining closer to their customers whilst still chasing cheaper labor. Construction ranks highly as a volume consumer of plastic products, but the bulk of this is extruded, e.g. sheet materials, piping, guttering and window profiles.

This is bound to be just an overview and you should do certainly do current research to get a feel for growth opportunities, requirements for suppliers, and other information pertinent to your business.

7.1.1 Where to look for information on particular markets

- Whole book and reports are devoted to individual markets, the companies within them and the products they make or buy. **Market researchers** like AMI (**www.amiplastics.com**) and others publish regular studies of markets for plastics processors. Look for those research companies which focus on just the sectors and regions which interest you, whose consultants will have real insight into market conditions (and may potentially offer personalized research if you choose to look more closely at specific opportunities).

- **Trade associations and other industry groups** often provide **directories** of their members (your potential customers) as well as information about industry structures and issues.

- **The internet** has also brought a whole new set of industry-specific information sources onto your desk, in the form of **digital magazines, networking groups, blogs, forums, wikis and others**.

- **Trade shows and industry conference**s are also a great way to get a feel for a new industry or sector. Walking the aisles or listening to the key speakers will give you a good feel for the shape and mood of an industry. Who are the key players and their suppliers, and where are they located? What issues are currently driving their strategies, and what new technologies are they showcasing or expecting from their suppliers. And finally, there is no substitute for personal contacts from these events when forging new business relationships.

7.2 Automotive/transportation

In 1973, the United States found itself with a gasoline (petrol) shortage and suddenly 'muscle' cars were out of vogue. The big tanks – Detroit boats, as they were called – guzzled too much gas, and suddenly America's love affair with the automobile found itself waning. So could we have the big, beautiful Ford, Chrysler, and Chevrolet models that we'd grown to love in the 1950s and 1960s in the US, or the big bold Mercedes, BMWs and Jaguars that Europe loves so much, yet lightweight them so that they consume less gasoline? The answer was a resounding "yes!" and plastics became part of that solution. Converting metal components to plastic didn't happen overnight in the automotive industry. But plastics eventually became the way to give passenger cars and light trucks the looks that consumers love, the lighter weight that reduced their fuel consumption, and the strength that these vehicles needed.

The plastics industry embraced the automotive industry and vice versa. It was love at first sight, and the US Upper Midwest – states like Michigan, Ohio, Illinois and Indiana – became a suppliers' heaven. Mold manufacturers and molders went after the automotive business because it was a huge industry, one that needed hundreds of thousands of parts, and one that would see growth eternal. As plastics materials evolved from what we consider commodity materials to the high-heat, chemical-resistant engineering thermoplastics, more applications evolved.

Roughly half of injection molded parts are for interior uses, and these have evolved from instrument panels, consoles, armrests, knee bolsters and rear parcel shelves as far as structural components such as 'B' pillars. **Weight reduction and aesthetics have also led to the replacement of exterior parts** such as bumpers, fenders, grilles and trim. Originally it was necessary to paint large plastic body parts to match the metal body color, but this process has now been overtaken by the use of film insert molding (FIM), which allows the molders to lay a color-matched film into the mold and inject the part behind it. FIM reduces pollution caused by painting, and reduces the capital costs involved in installing large paint booths containing spray equipment.

Engine and mechanical components make up around a quarter of injection molded parts, and some of the applications include air-intake manifolds, valve covers, pulleys, water pumps, air cleaners, fuel valves, engine and transmission covers, oil pans, fuel level devices, battery trays and many more. Temperature requirements for engine and mechanical parts continue to increase as air gaps are reduced and the proximity to heat sources increases, and resin producers have consequently extended the temperature performance of many of their grades.

One of those resin producers is Ticona (**www.ticona.com**), which produces a number of high-performance polymers such as Fortron polyphenylene sulfide (PPS), Celanex thermoplastic polyester (PBT), Celstran and Compel long fiber reinforced thermoplastics (LFRT), Vectra liquid crystal polymer (LCP), and Hostaform acetal copolymer (POM) which are found in many of today's vehicles.

Weight reduction, high temperature performance and chemical resistance have been key to the growth of plastics in automobiles but there are also other reasons for their increased usage:

- Plastics can contribute to stylish, contemporary designs with sleek curves and complex light clusters for example.
- Greater rigidity and tighter tolerances are possible with plastic moldings.
- Components can be redesigned to replace a complex assembly of metal parts with a single plastic structure.

- Moldings can incorporate features such as snap-fits, bosses, ribs and hinges, providing strength and simplifying assembly.

Overall, **replacing metal components with fewer engineered plastics parts can contribute to lower production costs by reducing inventory, assembly time and secondary operations, and improving quality control**. In the longer run 'lightweighting' will improve performance and fuel efficiency, and maintenance costs can also be reduced by the use of more durable, corrosion-resistant components.

However, with all this evolution into plastics came the complexity of being an automotive supplier. Custom injection molders and moldmakers were typically still relatively small companies, and while there was a lot of money to be made supplying the automotive industry with parts and assemblies through the OEMs and the tier one suppliers, the smaller, tier two suppliers, which include the molders and moldmakers, were thrown into a challenging new corporate world.

It was in the late 1980s that suppliers and the automotive industry became adversaries. Jose Ignacio Lopez de Arriortua came to General Motors as their vice president of global purchasing, and became known as the company's 'key cost-cutter.' It was Lopez who created the supplier model that refused down payment money on molds, postponed final payment on molds until sometime after PPAP (Production Part Approval Process), pushed payments to molders out by months, and generally made life extremely difficult for these small businesses.

For nearly two decades, automotive suppliers struggled to survive under these onerous terms. Large, tier one suppliers could manage, but the smaller tier two suppliers, including molders and moldmakers, had a more difficult time. One of the biggest problems was getting pass-through payments from the tier ones.

The economic recession which started around 2008 put a different spin on things in the automotive industry. Consumers stopped buying cars and sales tanked, and the suppliers began hurting. Some of the tier ones were already in bankruptcy or just beginning to emerge. Delphi (a major parts manufacturer) ended four years of bankruptcy in October 2009, emerging as a private company. Along that arduous path, this former unit of General Motors cut thousands of workers, divested several businesses and agreed to sell its steering systems operations and four plants back to GM. Today, Delphi is a much leaner more focused company whose core businesses are electronic and safety components, powertrain, thermal, electrical and electronic systems and replacement parts.

Smaller tier two suppliers, unable to sustain their business as their tier one customers fell onto hard times, also went out of business. However, this reduction in their supplier base has not been good for the tier ones, resulting in generally better terms for their surviving suppliers.

Maurice Sessel, Vice President Product Engineering for International Automotive Components (**www.iacgroup.com**), a global conglomerate of manufacturing companies that produce a range of components and assemblies for the automotive industry, gave the keynote address at the Society of Plastics Engineers Automotive TPO (thermoplastic Olefins) Conference in Detroit in 2009. His presentation, 'The Interior Challenge,' was optimistic as he shared real-world examples of how IAC has turned many of the challenges of 2009 into business opportunities. *"At IAC, we've used these variables (mounting industry, regulatory and economic challenges) into outlets of opportunity, rather than recoiling from them,"* said Sessel. He highlighted three key demands that will impact the future of the automotive industry:

- The unprecedented economic and industry times that include the lowest production volume in decades, reduced cash flow and increased prices.
- Demand from consumers for sustainable and natural materials, and transition to smaller, more fuel efficient vehicles.
- Tough government mandates for increased fuel efficiency and increased safety that continue to pressure automakers and their suppliers to reduce vehicle mass, while improving crash performance.

With all the focus on the plight of established auto manufacturers in the US and Europe and their suppliers, there has been little attention paid to China's automotive industry. However, this is certainly a business to be reckoned with. According to the AlixPartners 2010 China Auto Outlook, Chinese auto suppliers nearly doubled profit margins in 2009 due to record vehicle sales in China, although exports declined and materials costs rose. Overall continued growth is anticipated, along with mergers and acquisitions leading to more globally significant companies.[1]

At the time of writing it appears there are some signs that parts of the global auto industry are emerging from the recession, with considerable production increases seen in 2010, albeit partially driven by restocking. The challenge for carmakers and suppliers is always going to be how to further improve their competitiveness and grow their market share. Molders and moldmakers play a critical role in helping them to achieve this.

---

1 AlixPartners China Auto Outlook, **2010**, www.alixpartners.com.

> ☞ **While the opportunities in the automotive sector are extensive and ongoing for those who've entrenched themselves in this market, it's a difficult market to enter.**

Competitive pressures in the automotive industry always include China. Many mold manufacturers and molders for the automotive industry continue to quote against China at the behest of purchasing engineers at the OEMs. In one instance, a major OEM announced in 2009 that it would release some $24 million in tooling to Michigan moldmakers, but in making that announcement to much applause, neglected to say that they were requiring 70% of their tooling to be obtained from a low-cost country (LCC).

To be a profitable supplier to the automotive industry, custom injection molders must have a number of competitive advantages:

- Cost advantages – reducing costs-to-manufacture is critical to the automotive industry.
- Quality is a must – automotive OEMs demand quality, and while quality and cost are often at odds, to be an automotive supplier both are essential.
- A good financial position – only the strong can survive the terms and conditions of the automotive industry. Molders need a financial position that will allow them to weather the long payment terms that are still demanded. While some suppliers have successfully negotiated better payment terms, molders for this industry still require a strong cash position and good lines of credit.
- An ability to produce molds in China. In many instances, partnering with a reliable Chinese moldmaker is the only way to capture good molding business.

While the opportunities in the automotive sector are extensive and ongoing for those who've entrenched themselves in this market, it's a difficult market to enter. **Many of the automotive OEMs have pared back their supplier base** from literally hundreds of suppliers in the 1980s and early 1990s, to fewer than 100 today in an effort to reduce the complexity of the supply chain and to maintain tighter control. The biggest issue for suppliers in this industry is learning to live with the automotive industry's payment terms and conditions, and working as a middle-man between the OEM and the moldmaker to ensure the necessary quality and price point. Molders in this arena also need a good understanding of the processing of the high-performance polymers.

Case histories: New materials and innovations

Ticona offers many engineering thermoplastics for use in automotive, particularly for under-hood applications and fuel systems components such as its Fortron 6162XF, a mineral/glass-reinforced polyphenylene sulfide (PPS). The new advanced fuel resistant Fortron PPS grade has reduced weight gain by more than 45% compared with other Fortron grades, and these materials are well suited for components that need extremely high dimensional stability during storage in aggressive fuels, including impellers, outlet covers and fuel injection components.

Fortron PPS is versatile and can withstand long-term exposure to gasoline, diesel, bio-diesel and aggressive fuels containing oxygenates such as methanol and ethanol, as well as to natural gas and other fuels in alternative propulsion systems. It also offers a high continuous use temperature of 160°C (320°F) to 240°C (464°F), and short-term temperature resistance up to 270°C (518°F). Additionally, it offers a potential weight savings of up to 50% versus metal.

SABIC Innovative Plastics (**www.sabic-ip.com**) introduced several new applications for its materials recently including a newly-designed injection molded steering wheel molded from Lexan EXL copolymer resin. The steering wheel features a two-part molded design that offers an alternative to the typical steering wheel comprised of a die-cast magnesium or aluminum armature overmolded with flexible urethane foam.

Lexan EXL glass-filled polycarbonate-siloxane copolymer resins gives stiffness and a high degree of impact strength over a wide temperature range. Sabic developed two designs: a two-part injection molded design with a leather wrap as a high-end solution and a one-piece injection-molded armature with a PU overmolding. Both are attached to the steering column with a small metal hub.

Benefits include up to a 20% system cost reduction, and a 40% reduction in mass compared to a magnesium or aluminum alloy steering wheel. It also offers greater design flexibility, and new styling options such as special colors and effects. Simpler manufacturing is made possible by reducing the number of secondary processes required with previous steering wheels.

Hella (**www.hella.com**), a tier one auto lighting supplier in York, South Carolina, selected Sabic's XHT high-heat PC resins to create a metalized headlamp bezel and a clear, ice blue-tinted bezel for two major European car manufacturers. "*As temperatures climb with new headlamp technology, many traditional materials can't handle the heat,*" noted Dr. Markus Richter, spokesman for Hella.

PolyOne (**www.polyone.com**) helped Ford Motor Co. provide additional eco-friendly features to its Escape Hybrid, by focusing on the polypropylene engine cover, which had been coated with metallic paint to achieve a high-quality finish. Ford designers wanted to eliminate painting and its associated VOC emissions without losing the aesthetic appeal of a metallic look.

Ford liked PolyOne's Maxxam FX Metal, a metallic-look, talc-filled PP compound, so they asked supplier MPC Inc. (**www.mpc-inc.com**) to evaluate the material in the engine cover application. PolyOne worked with the mold supplier to design the injection mold to eliminate visible knit lines by routing material flow in the mold so that flow fronts meet at the edges of the cover. The Maxxam FX Metal met all of Ford's performance requirements including impact strength, heat deflection temperature and tensile strength while also providing the aesthetic requirements of the engine cover.

On the interiors front, PolyOne helped a German automotive OEM develop a new application involving Onflex-S TPE with UV stabilization. The OEM had developed interior TPE parts in the past, the engineers requested help in creating an elastomeric insert of the map case pocket used in each door of a compact SUV now in series production. PolyOne's Onflex-S with UV stabilization provided improved weathering performance, and helped the OEM reduce cycle times due to higher flow rates than other SBC-based TPEs.

7.3 Medical devices/disposables/durables

Medical device manufacturing has long been viewed as a highly desirable business for the plastics molder, offering good profits for those companies able to deliver high quality precision-molded products. **The market is generally considered to be largely recession-proof, with an apparently unstoppable demand driven by an expanding but ageing global population and research-driven technological advances.** Historically the sector has also been resistant to potentially compromising quality by sourcing overseas, although this is no longer true across the board.

According to a report from The Freedonia Group,[2] US demand for disposable medical supplies will increase 4.6% annually to $59 billion in 2013, with the best growth opportunities anticipated in intermittent catheters for urinary drainage, dry powder inhalers, prefilled syringes and transdermal patches for drug delivery, blood glucose test strips for diabetes monitoring, and daily contact lenses. Medical instruments and devices were anticipated to grow from $13.5 billion in 2008 (6.0% annual growth from 2003-2008) to $17.5 billion in 2013 (5.4% growth from 2008-2013).

2 Disposable Medical Supplies to 2013 - Market Research, Market Share, Market Size, Sales, Demand Forecast, Market Leaders, Company Profiles, Industry Trends, study 2476, July **2009**, www.freedoniagroup.com.

The US industry is analysed further in a study published by Ernst and Young,[3] which observes that US life expectancy has reached an all-time high (currently 78.1 years), and the annual death rate has dropped to an all-time low of 760 deaths per 100,000 people. The medical technology industry also contributes significantly to the US economy, providing jobs that pay on average 30% more than the average US job. Due to a 'multiplier effect,' every medtech job also results in 4.5 additional jobs across the nation. The report also demonstrates that the industry has been competitive, as medical device price changes have been historically low, lagging behind both the Consumer Price Index and the Medical Services Price index from 1989 to 2006.

> ☞ ***"New products are often difficult to get into the marketplace because of lengthy and costly FDA protocols, which means many mature products can't readily change significantly. FDA approvals for medical products are in some respects like PPAP in the automotive industry in that it can take anywhere from six to eight months or longer, to gain approval."***

At the time of writing this book reform of health-care funding was causing some uncertainty in the US, alongside the global recession, but nonetheless the industry remains stable and innovative.

"*The industry's financial performance . . . has largely held steady,*" notes the Ernst and Young report. "*The revenues of publicly traded medtech companies in the US and Europe grew 11% to US$289 billion, and while net income fell by 11% in 2008, this was mainly due to large one-time charges.*" In the first half of 2009 the author reported that revenues remained essentially flat (a decline of less than 1%) compared to the same period in 2008.

According to the Ernst & Young report, the US has 969 public (including conglomerates) and venture-backed medtechs, and of those 52% were headquartered in one of three states – California (186 VC-backed companies and 33 public companies), Massachusetts (82 VC-backed companies and 41 public companies), and Minnesota (66 VC-backed companies and 39 public companies). Perhaps more surprisingly, similar to the geographic concentration within the US, 56% of the 738 public and venture backed European medtechs are headquartered in just three countries: the UK (138 VC-backed companies and 39 public companies), Israel (115 VC-backed companies and 25 public companies), and Germany (101 VC-backed companies and 29 public companies).

---

3 Ubl S.J., Pulse of the Industry Medical Technology Report, Ernst & Young, November **2009**, www.ey.com.

(Note these figures only refer to headquarters, and not manufacturing locations which are more widely spread).

There are certainly opportunities in the medical market for molders with clean room/clean environment molding capabilities, and approved good manufacturing practices (GMPs) in place. However, molders and moldmakers wanting to specialize in supplying medical device OEMs must also be aware of the complex processes these OEMs go through to gain approval for new or improved products.

☞ **People will always need medical products.**

How custom molders and moldmakers can best serve the medical device industry was covered in an article by this author in 2010.[4] *"Medical OEMs are looking to drive down costs while driving up profits to satisfy stakeholders,"* said Jerry Seidelman, sales and marketing manager for Tech Mold Inc. (**www.techmold.com**), a mold manufacturer working primarily with the medical and pharmaceutical industries. *"New products are often difficult to get into the marketplace because of lengthy and costly FDA protocols, which means many mature products can't readily change significantly. FDA approvals for medical products are in some respects like PPAP in the automotive industry in that it can take anywhere from six to eight months or longer, to gain approval."* The question of gaining approval can cause difficulties for molders unfamiliar with the medical business. They may devote considerable time and resources to developing a product or process alongside an OEM, only to find that it is not approved, and the expected ongoing manufacturing contract never materializes.

Seidelman also noted that medical device OEMs expect much more from their suppliers than they did a decade ago. *"Medical device OEMs expect their mold manufacturers to have a range of capabilities that might include collaborative engineering, new product development assistance, prototype or pre-production mold capabilities, the ability to design molds for automation and even integrate complete manufacturing cells for optimum molding in a clean room environment."*

As mentioned above, one reason for the entrance of any molders and moldmakers into the medical industry segment is the belief that it's recession-proof – "People will always need medical products." Another reason so many molders and moldmakers want to get into the medical industry is the perception that this is the one industry that will never go offshore. This belief is encouraged by the level of regulation and the requirements for good manufacturing practices (GMPs). However, neither of those reasons is completely true.

---

4 Goldsberry C., Market Snapshot: Medical devices, February 24, **2010**, www.plasticstoday.com.

The medical industry in China represents a huge opportunity for US device OEMs, which makes it tempting for them to take their manufacturing there. According to a study from Espicom, growth in the Chinese medical device market for 2010 was estimated at 13.6%, due to the vast population and government commitment to developing the health-care system.[5] Some western molders have put part of their manufacturing in China – typically those products that will be sold in the Chinese market – and found some level of success. Others have tried it for products that are exported to the US market and found problems with that model, but many still believe in the 'China solution' to their manufacturing needs.

Bill Kushmaul, president & CEO of Tech Mold Inc., believes that most in the medical industry haven't learned that lesson. *"Good, bad or indifferent to China – they all say 'I need a China solution.' They believe any good strategy requires a China solution,"* Kushmaul states. *"Even big medical companies are seeking a China solution. Years ago we used to say 'medical will never be done in China.' We know today that's not true."*

China (in 2010) was in the process of implementing its own rules with regard to medical device manufacturing to bring itself in line with US FDA regulations. Beginning early in 2006 the State Food and Drug Administration (SFDA) of the People's Republic of China, the regulatory body in charge of the manufacturing, distribution and use of medical devices, attempted to implement good manufacturing practices (GMPs) for medical device products nationwide, according to a report by law firm Squire Sanders & Dempsey LLP.[6] *"By raising the standards for medical device manufacturing, the implementation of GMP was expected to benefit the health of the general public in China and increase the competitiveness of China-based medical device manufacturers in the global market,"* said the report. There was, however, no general GMP rule applicable to all kinds of medical devices until December 2009, when the SFDA issued a package of regulations addressing GMP for medical devices including the Regulation on Medical Device GMP (GMP Rules) and the Regulation on Examination of Medical Device GMP (GMP Exam Rules). Both of these rules were to become effective on January 1, 2011, leaving one year for medical device manufacturers to make improvements.[7]

---

5 The Medical Device Market: China Opportunities and Challenges, February 10, **2011**, www.espicom.com.

6 PRC medical device good manufacturing practice rules issued, Squire Sanders & Dempsey LLP, Lexology of the Association of Corporate Counsel, January 22, **2010**.

7 Goldsberry C., Medical OEM struggles with China manufacturing, February 1, **2010**, www.plasticstoday.com.

7.3.1 European outlook for medical molding

While medical molding seemed so safe for both European and North American molders serving this market segment, outsourcing has had an impact on medical device manufacturing. A report from Frost & Sullivan as far back as 2005[8] noted that Asia is emerging as a "*prime low-cost medical devices manufacturing spot,*" and this is of concern to European medical device companies that "*run the risk of losing valuable market share.*" While this growing trend presents an obvious threat to European medical manufacturing companies, some view it as an opportunity to expand into the vast Asian market. However, noted Frost & Sullivan, these companies "*will have to move fast to capitalize on early growth opportunities.*"

According to the report, many European medical device manufacturing companies are beginning to establish their own production facilities in India and China, the two most populous countries that are emerging as the 'favored destinations' for European companies. The companies are realizing the reduced operations costs, particularly in labor. There are other costs however, that cut into the profits that come from manufacturing there such as the costs of distribution – shipping products back to the European market, and some companies are partnering with contract manufacturers that already have established manufacturing facilities in those countries, and outsource their work to these. *"Shifting manufacturing facilities to Asia could also promote better penetration of the Asia medical devices market,"* according to Claire Staniforth, Frost & Sullivan Industry Analyst.

The trend toward outsourcing production operations from Europe to Asia is certainly upwards as these countries are now focussing of quality and conforming with international standards. Meeting these quality manufacturing standards also helps ease any concerns that health-care providers in the European market may have about the products being inferior or second-rate merely because they are less expensive. With health-care providers constantly implementing cost-containment measures, purchasing less expensive medical devices that nevertheless meet certain quality requirements may prove an attractive option, according to Frost & Sullivan.

"When assessing the attractiveness of China and India for manufacturing of medical devices, China has a much higher relative appeal as investment in infrastructure and global foreign direct investment are high," added Ms. Staniforth. *"However, there is a greater availability of skilled labour in India, which could sway companies into investing in manufacture there instead."* Some European medical device manufacturers are reported to be choosing partnerships or strategic alliances

8 Impact of Low Cost Asian Manufacturing on the European Medical Devices Markets, December 7, **2005**, www.frost.com.

with non-Asian contract medical device manufacturers. These companies are more likely to have a strong supply chain in place and thereby have better negotiating power.

However Frost & Sullivan suggest that the biggest challenge European companies are likely to face is with regard to intellectual property (IP) protection. That presents a significant risk to medical device companies, as there are thousands of 'back door' manufacturing plants that copy products, mainly for sale in China and India, although many find their way back to Europe where they are sold at much reduced prices.

"European companies must therefore ensure that they conduct business only with manufacturing organizations that understand and appreciate the importance of IP protection," said Ms. Staniforth. *"In Asia, contract research organizations (CROs) – essentially research and development outsourcing companies – are starting to introduce IP standards to increase confidence levels within the foreign investment community as well as to promote greater outsourcing interest in both R&D and manufacturing in Asia."*

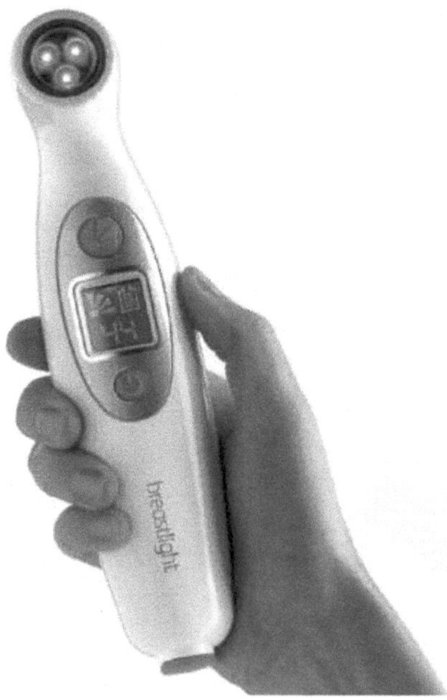

Figure 7.1

Innovation in medical instrumentation, such as this Breast Light that allows women to perform breast self-exams more thoroughly, offers new opportunities for molders (*Sabic Innovative Plastics)*

7.4 Packaging – food, cosmetic, and consumer

> ☞ ***"Consumers are seeking 'grab and go' convenience at retail, and they're also increasingly looking to packaging for portion control."***

Plastic packaging is ubiquitous, and this is another area which is known for high-volume injection molding typically using extremely large, multi-cavity molds, particularly for components such as caps and closures, small bottles and vials, and preforms for injection-blow molded water bottles. Other molded applications include specialty packaging for pharmaceuticals (vials and caps); medical products; and cosmetic applications such as intricate, fancy caps for perfumes, and compacts for eye shadow, cheek blush and face powder.

BMO Capital Markets (**www.bmocm.com**) track the packaging sector, and noted falling demand in their December 2009 publication, The Converter. M&A activity in the packaging sector had been heavy over the past few years, however BMO's 2009 report released in The Converter (January 2010) also noted that this declined sharply in 2008 and 2009. There were just 169 transactions announced globally in this sector in 2009, which represented a 33% decline versus the 251 transactions announced in 2008 and a 56% decline from the 2007 peak. North American deal volume also declined, with 68 announced transactions in 2009, a decline of 20% from the 85 deals announced in 2008.

Whilst M&A activity has slowed down, this is certainly not stopping innovation in the packaging industry. According to a press release from the Packaging Machinery Manufacturers Institute (**www.pmmi.org**), the beverage packaging industry is experiencing a period of innovation in design (packaging shapes, sizes and features), materials and efficiency intended to meet cost and environmental concerns while enticing splintering consumer groups.

"Single-serve containers are particularly strong market drivers. Consumers are seeking 'grab and go' convenience at retail, and they're also increasingly looking to packaging for portion control – even with beverages," said Charles D. Yuska, president & CEO of PMMI. *"Sustainability is also a market driver as consumers expect brands to play a role in helping the environment."*

Some trends that PMMI expects to impact the current market include the following:

- Eye-catching single serve packaging,
- Cost-effective 'green' packaging, driven by manufacturers' cost and environmental concerns,

- Plastics as the 'material of choice' because of rapid innovations leading to lighter, more eco-friendly packaging,
- New packaging machinery that is flexible and supports the latest container technologies,
- Rapid growth for pouch containers, as manufacturing technologies bring costs and speeds of pouch beverage production in line with high-speed glass bottle production,
- Minimizing secondary packaging to cut costs and reduce overall carbon footprints.

According to Yuska, flexible packaging is leading the response to changing consumer demands, which could be a threat to the injection molded container segment as pouches catch on. Already, many food containers, especially take-out containers, have embraced thermoforming technology – an interrupter to injection molding – with high-speed, high-volume forming machines moving into the packaging market segment.

Overall the good news from PMMI is that the $22 billion US market for beverage primary packaging is dominated by plastic with 40% of consumer goods companies using a variety of formats to attract consumers. Thus there are many opportunities for injection molders in this arena, including technology such as **in-mold labeling**, which has been big in Europe for at least a decade, and is finally catching on in the United States.

It was reported at a 2009 conference,[9] that the use of in-mold labels for packaging is seeing some additional interest in the United States, but is really taking off in China and India. According to William Llewellyn, VP and senior consultant at AWA Alexander Watson Associates, the global label market saw demand of 39.2 billion m^2 in 2008, or 1.5-2% growth compared to 2007. This growth was far less than the 5-7% seen in 2007, however he also noted that the industry was seeing some signs of revival, and 2009 could end up better that 2008 when all the numbers come in. Of global demand for labels, 26% of that was in the North American market with 10.5 billion m^2; 30% of the demand was from Europe; 31% came from Asia with most of that coming from China; and 9% from South America. India represented about 16% of the Asian market demand. *"China and India are really taking over the market for IML,"* noted Llewellyn.

The reason, apparently, is that the perception of IML's extra cost often isn't borne out in reality, according to the In-Mold Decorating Association

9 IMLCON '09,Oct 28-30, Phoenix, AZ, **2009**, Alexander Watson Associates, www.awa-bv.com.

(**www.imdassociation.com**). When the labor and extra steps involved in secondary operations for applying glue-backed labels or pressure-sensitive (PS) labels are considered, then IML can compete on a cost basis, according to IMDA.

US injection molders for the packaging market haven't entered the in-mold labeling technology in any big way because their OEM customers – the Fortune 100 companies – are reported to be reluctant to go that route. In North America the extrusion blow industry is the largest user of in-mold labels (88% of the label market). Domestic supply of labels also limits the growth rate in North America as most of the labels are made in Europe, and North America is currently importing some 50% of the labels it uses. Peter Paelinck, North American business development manager for Belgian label supplier Verstraete (**www.verstraete.be**), agreed with Llewellyn's analysis. *"That 50% import figure of labels to North America is about right,"* Paelinck said. *"Approximately 15% of the 28 million labels Verstraete makes come to North America. People underestimate the potential of the IML market in North America."*

For custom molders serving the packaging market and wanting to enter the IML technology sector, it requires molds specially built for the technology, and robotics to grab the label, put it into the mold cavity and take the parts out, as well as systems to check the label is on correctly, and move the product down the line. All of this needs to be done at high-speed to accommodate the volumes. Molders must work with the moldmaker, the robotic OEM, the auxiliary equipment OEM for conveyor systems, and the label manufacturer, and this needs to be done up-front so that it all comes together in the molding cell. This requires a lot of capital investment, which many molders have been unwilling to commit to date, perhaps until they see more enthusiasm from the OEMs. The reluctance of the OEMs may be due in part to the high number of different stock items which are required across a product range, and a loss of some flexibility in manufacturing. Because IML molds the labels into the package, the labels are permanent. Should marketing change something about the product, or want a new label, all the containers that have been labeled with IML have to be discarded (or recycled if the label is of the same plastic material as the container). However, opportunities certainly exist for molders who make the investment in this technology for their OEM customers, and this has been especially true in Europe, where IML has a longer history and where most of the labels for IML are currently produced.

In general, most OEMs in the packaging industry have in-house or proprietary molding operations, which doesn't leave a lot of room for custom injection molders to enter this market and capture the routine work. There are some smaller opportunities in this segment such as trial or initial runs for a new product, particularly in a mold build project, and smaller companies that have a specialty 'niche' packaging application often use outside molding suppliers. In addition, many OEMs will turn to molding and moldmaking suppliers where they offer a specific capability or niche in which the OEM doesn't want to invest.

Case histories: New products and solutions

Finding ways to add value for their customers is a primary goal for many molders and moldmakers. One company, R&D/Leverage (**www.rdleverage.com**) with facilities in Missouri, US and Nottinghamshire, U.K., has led the way in adding value for its customers in the food, beverage, health-care, home and personal care industries, with packaging solutions to support the branded products, including concept development, design, model making and prototyping, mold manufacturing, and contract molding. R&D Integrated Solutions in Plastics is a full-service plastic product solutions company, and Leverage Integrated Industrial Design is their full service consumer packaging design component, located in a separate facility within the R&D campus. The company provides a 'one-stop-shop' approach, offering a complete package which begins with consumer research to pre-validate conceptual designs, then extends to retail audits, prototypes, unit tool development and subsequently to full production molds.

Rexam PLC (**www.rexam.com**), a global consumer products packaging company, also uses molding technology to offer innovative solutions such as their new Nea pump and the Prodigio airless dispenser. Prodigio is designed for many applications such as facial care products (moisturizing creams, anti-aging creams, masks or facial scrubs). A patented mechanism prevents the system from clogging with dried cream in the dispenser. Made of polypropylene and polyethylene with a single capacity (50 ml, dispensed in 1 ml doses), Prodigio is 100% recyclable. Nea pumps are available in plastic or metal, with customizable components (cap, push-button and collar available in a large choice of colors and finishes), and are available in three sizes (100ml, 150ml, and 250ml).

This type of sophisticated and stylish package helps Rexam's customers in the business of personal care products to meet the demands of their consumers. Typically these are driven by environmental concerns, such as reduced weight and recyclability, and general avoidance of excessive packaging. Customers also want packs which can be completely emptied, avoiding waste. Where a package has several components, use of materials which can be recycled together is an important consideration, and Rexam have also noted that in-mold labeling (IML) is becoming increasingly important to the personal care industry for this reason.

Another area of innovation identified by Rexam is the small trial size package which meets a number of consumer needs in personal care products. Customers may wish to try a product before purchasing a larger size, but in emerging markets trial sizes are popular because customers cannot necessarily afford the full-sized product. Trial sizes can also enable customers to purchase 'luxury' products.[10]

Figure 7.2

Cosmetic packaging requires good aesthetics and designs that appeal to consumers like this Nea pump top (*Rexam)*

Typically consumer products companies are large global concerns, and hence the moldmakers and molders which serve them must also be large and global. (Another well-known example is Nypro Corp., a large global custom molder that serves large OEMs worldwide.) Smaller companies wanting to break into the packaging business - both molders and moldmakers – might look to serve these larger molders and moldmakers in a 'tier two' capacity, much like in the automotive industry.

Specialization is obviously an advantage, and in some cases clean room molding areas are required, although many containers, particularly for food, undergo

10 Goldsberry C., Market Snapshot - Personal Care – Innovation in Packaging, December, **2009**, www.plasticstoday.com.

a sterilization process prior to fill. Equally importantly, many food packaging processors are located near the food processors, so that the distance between the container producer and the fill plant is relatively short to reduce shipping costs. Typically food packaging in the USA requires the use of FDA food-contact approved plastics. Molders who serve this industry must generally follow the FDA guidelines for good manufacturing practices (GMPs).

7.5 Industrial

Industrial products are often hidden within larger assemblies, and this lack of visibility might be the reason that few molders ever think of the industrial sector as an important market from which to capture new business. Yet it is a growing market and players surveyed in the industrial sector have reported a positive outlook. PricewaterhouseCoopers survey 60 senior industrial executives in the USA for their quarterly Manufacturing Barometer report.[11] The January 2011 edition showed increasing optimism about the US economy and company growth rates, growing overseas sales, stability for jobs and other positive indicators.

Figure 7.3

Industrial components are often hidden within the products themselves, like this Stanyl water pump component *(DSM Engineering Plastics)*

11 Manufacturing Barometer, 1Q, **2011**, PricewaterhouseCoopers, www.pwc.com/us/en/industrial-manufacturing/barometer-manufacturing.

Typically, industrial components have to be rugged, tough and able to withstand chemicals, abrasive wear, and harsh environments, which means that many components are thick-walled parts molded using engineering thermoplastics such as polyetheretherketone (PEEK) and nylons among others. Many industrial parts that used to be made of metal are now being molded in engineering thermoplastics to take advantage of the characteristics that plastic offers over metal, such as corrosion and heat resistance. For example, Sabic Innovative Plastics offers high-performance Noryl (modified polyphenylene ether), Ultem (polyetherimide) and Cycoloy (polycarbonate/ABS blend) resins to replace traditional metals such as copper and brass used in condenser housings and manifolds for wall-mounted boilers.

While industrial products can theoretically be accommodated by almost any molder, you should have the capability and expertise to mold extremely thick-walled parts using heavy-duty engineering thermoplastics such as Ultem and PEEK, that are typically molded at extremely high temperatures. (These have taken the place of thermoset materials in many applications). Good quality control procedures are frequently required for these products, many of which are used in critical industries such as chemical plant pumps, natural gas pipelines, water and waste-water pumping stations.

7.6 Aerospace/defense

The aerospace and defense industries offer some good opportunities for those molders who have the necessary military certifications to be a supplier to the US government for example, and to the large aircraft manufacturers, such as MIL-STD (961E, 962D and 967) and MIL-SPEC, which can be obtained from the US Department of Defense. Opportunities for molding components for military and defense applications can be found online once you are certified. Europe offers similar MIL-SPEC certifications such as EUROCAE (European Organization for Civil Aviation Equipment). A good website for more information is **www.everyspec.com.** Like the industrial market segment, the aerospace and defense segment generally demands rigid quality control, and uses materials on the high-end of the engineering resin spectrum, including heavily-filled compounds. The aerospace/aircraft segment also uses composite materials – generally outside the spectrum of the custom injection molding operation.

This is a narrow market, with jobs typically only offered and open to certified suppliers, so these companies tend to specialize, much like those in the automotive sector.

Case history: Tri-Mack Plastics Manufacturing Corporation

Tri-Mack Plastics Manufacturing Corporation (**www.trimack.com**), a leading manufacturer of high-performance engineered components primarily for the aircraft industry, received General Electric Aviation's Supplier Quality Awareness Award in November 2009. Tri-Mack was the first supplier selected for the award, which recognizes high standards for productivity and quality.

Tri-Mack had been working closely with GE Aviation (formerly GE Aircraft Engines) for more than 30 years before gaining the award in recognition of Tri-Mack's services and achievements in supplying parts for the GENx-1b, a next-generation turbofan engine. The GENx-1b incorporates a new technology for oil cooling that uses lightweight composite fairings, for which Tri-Mack manufactured six different injection-molded parts in extremely complex mold tooling on a very accelerated schedule. Tri-Mack's engineering staff worked concurrently with GE on the GENx-1b to optimize part designs for producibility and overall cost-effectiveness. Their tool design for the project used proprietary hand loads to mold complex geometry, including molded in-place metal inserts. The company also employed advanced selective laser sintering technology in its strategy to save time in the development process.

Tri-Mack actually began life as a compression molder, which led the company into injection molding using high-temperature materials such as polyamide and polyimide and highly filled materials, and subsequently making their own tooling for these specialised materials. Like most other markets, President Ed Mack observes that there is pressure to the customer's requirements with zero-defects at reduced costs (3-6% every year) and reduced lead times. "*It's hard to talk about quality unless you talk about price and delivery as well,*" Mack noted. "*Quality is so woven into the culture of this company, it's hard to separate it and just talk about parts. Most of the processes are covered by AS 9200 (a MIL-SPEC for aircraft manufacturing) – these are used to deliver price reductions and on-time delivery.*"[12]

7.7 Appliances

The white goods market has seen a lot of consolidation over the past few years, with Whirlpool coming out a big winner. With Whirlpool's acquisition of Maytag, it is now one of the largest appliance makers in the world, given that Maytag had acquired some smaller players (Amana, for one) prior to Whirlpool's acquisition. GE Appliances is another major player in the industry, along with others including Indesit and Electrolux. However, as in most other segments there has been a

12 Goldsberry C., As I See It: An Evolution in Flight, March 1, **2010**, www.plasticstoday.com.

trend towards globalization with the growth of Far Eastern manufacturers such as Samsung and LG. While the appliance industry is a large industry, it was badly affected by the decline in the housing market in the US and Europe, since big sales come from builders making bulk purchases for large housing developments. This is particularly true in the US where we typically have much larger appliances, and the housing boom of the early 2000s really provided huge opportunities for molders that specialize in this market.

Some sales will still come from homeowners upgrading their older appliances to new ones that have more features and improved energy efficiency. There are even smart appliances starting to appear on the market now which may also encourage buyers to upgrade. Before the recent economic downturn there was certainly a trend amongst appliance manufacturers away from the more competition-sensitive, value-priced products to the high-end feature-filled appliance models to improve margins.

While there used to be dozens of appliance makers with plants across the United States, many plants have closed as these companies consolidated, and much of that work moved to Mexico. (Asia is a less attractive outsourcing region for major appliances because of the distance and shipping costs to the United States.) European appliance manufacturing has tended to move to lower cost countries such as Turkey, although some European appliance makers are in Mexico, such as Electrolux and Bosch.

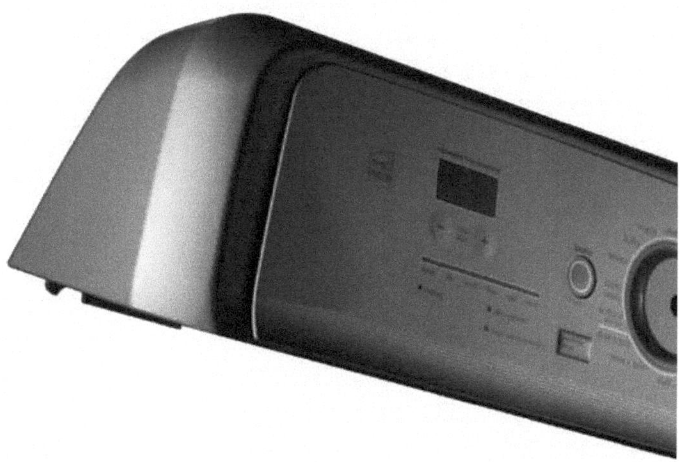

Figure 7.4

Geon FX Metal Vinyl gave an appliance maker the option to design and mold this washer/ dryer console in one piece, consolidating four parts and eliminating painting *(Polyone)*

Jones Plastics (**www.jonesplastic.com**), a multi-plant custom injection molding company that does a considerable amount of business with the appliance industry, reported a 10% fall in this part of its business in 2008, having previously observed good demand for appliances for 'intuitive' features – electronic bells and whistles – before the global economic downturn.

There are also some opportunities for molders in the small appliance market, but not as many as there used to be. Much of that molding has moved to China and Mexico – primarily China. There has also been some consolidation in that market, for example Jarden Corp. is a very diverse company that has purchased numerous small appliance makers such as Sunbeam, Oster, Mr. Coffee, Crock-Pot, Health-o-meter, Rival and FoodSaver and others.[13]

7.8 Toys, recreational and sporting goods

Back in the 1950s and 1960s, when plastics became prevalent in toys, many molders in the Southern California region were molders and moldmakers to one of the biggest names in toys – Mattel Corp. In fact, it was only about two decades ago that most of the major toy manufacturers had plants in the US and Europe. For the US toy companies, Mexico was the first big draw as a way to lower costs, but that lasted barely five years. Then the toy makers began the move to offshore, to Asia primarily. While it may have resulted in reduced costs-to-manufacture for the toy makers, it didn't reduce the price of toys to the consumer (and nor did it do anything for the safety standards of many of these toys). In the last couple of years there have been significant safety questions around Chinese imports (45 million toys were recalled for various reasons including the presence of heavy metals and lead paints), leading in the US to the Consumer Product Safety Improvement Act (February 2010), and there is certainly hope that some of this molding work may return to the US and Europe. At the moment however, opportunities for custom molders are still limited.

Sporting goods and recreational markets offer good opportunities because no matter how bad the economy get, people always want to find ways to enjoy themselves. According to the Sporting Goods Manufacturers Association (**www.sgma.com**), the US sporting goods industry was a $72 billion business (at wholesale) in 2009, a 4.3% drop from 2008, primarily due to the poor economy. However, sales for 2010 were expected to increase by 2% and 4.5%. Molders for sporting goods and recreational products typically have expertise with engineering grade thermoplastics. Piaggio (**www.piaggio.com**), Italian maker

13 Clare Goldsberry, Market Snapshot - Appliances, February, **2009**, www.plasticstoday.com.

of motorcycles, reportedly chose DuPont's Zytel HTN for its liquid coolant Low Emission Advanced Engine Range (LEADER), for its better sound- and vibration-dampening characteristics. Ticona Engineering Polymers finds that sporting goods and recreation offer ideal applications for its GUR ultrahigh-molecular-weight polyethylene, Hostaform. GUR-UHMW-PE is used in such applications as chain track wheel for snowmobiles, as well as the sliding rail for inline skates. Custom molders who serve this market are hopeful about the future, and have said that many of their customers are planning new models in their snowmobile and jet-ski lines, and new product innovations are on the drawing boards.

There are some certainly niche opportunities for molders in the toy and leisure market, primarily with entrepreneurs who are afraid to take their new toy idea to China for reasons including fear of intellectual property theft, shipping costs and general lack of control.

Figure 7.5

These discs are molded for Gateway Disc Sports for its new disc golf game *(RTP Co.)*

7.9 Building and construction

The construction industry has seen a major downturn in the current recession, both in residential and commercial building, and even the home improvement sector has struggled (although some people will choose to improve their homes when moving house is more difficult).

The largest volumes of plastics used in this sector are extruded (gutters, pipes, window profiles, etc.), although there are a many more injection molded products than there used to be, including injection molded siding panels that are replacing extruded vinyl siding; solar shingles to provide energy to homes; and roofing shingles designed to look exactly like real slate and shake shingles but offering improved durability or reduced cost.

For example, DaVinci Roofscapes (**www.davinciroofscapes.com**) manufactures injection molded roofing tiles, including snap-fit, self-locating tiles suitable for use in hurricane zones. PlyGem (**www.plygem.com**) manufactures a variety of polymer siding products, and say that their injection molded Cedar Shake and Shingle products offer a natural look, using variegated colors and molded finishes created by digitizing the texture of real wood.

Figure 7.6

These EcoBlend roofing tiles are injection molded using a blend of virgin and regrind to produce a strong, long-lasting roof *(DaVinci Roofscapes)*

While many building products companies do their own molding, there is some opportunity for custom molders in smaller types of products in this market such as light switches/panels, lighting products, and decorative items. And some of the products that were sent offshore to China have come back. Hy-Lite Blocks (**www.hy-lite.com**), owned by US Block Windows since March of 2009, molds architectural acrylic blocks for use in privacy windows, indoor radius walls and partitions. The company recently moved their molding back to its US manufacturing facility in Pensacola, FL, from China citing shipping costs as a key reason.[14]

7.10 Electrical/electronic components

Electrical and electronic components span a number of markets including automotive, computers, telecommunications, industrial controls and more. Most of the electronics equipment we all have in our home today, particularly consumer electronics such as television sets, stereo components, PDA's, gaming consoles and other consumer products, are made in Asia - China in particular. However there are still opportunities for molders to make electrical and electronic components which are used in other industry sectors. Electrical and electronic components offer molders opportunities for high-end business using engineering thermoplastics in articles that were often once compression molded. Newer thermoplastic materials and compounds provide optimal properties such as anti-static and thermal conductivity, and also the potential for recyclability. Molders in this business are likely to have the ability to mold parts holding extremely tight tolerances, especially in products such as connector components.

7.10.1 Automotive applications

E&E has been a major driver in the development of innovative new materials in automotive applications, according to Richard Frissen at DSM Engineering Plastics (**www.dsm.com**). *"For years, the value of electric and electronics content in automotive has been growing between 5-10% per annum. New electrical systems and components are added to improve comfort, safety or fuel economy. Engineering plastics play an important role in this segment, mostly as casings to contain and protect electrical and electronic parts from their environment and/or to secure electrical insulation."*

---

14 Goldsberry C., Market Snapshot - Building & Construction, November, **2009**, www.plasticstoday.com.

"Typical electrical applications in automotives include components such as junction boxes, ECU (electronic control unit) housings, E-motors, gear housings, bobbins, connectors, switches and wire harness insulation. Another main driver is the hybridization and advanced battery technology as the most predominant 'green' trend of the industry, with its main goal to realize fuel economy and reduced CO_2 emissions."

With the move towards hybrid and electric vehicles, electrically-driven components are replacing mechanical components in areas such as electrical power steering, electrical water and oil pumps. According to Frissen, *"by replacing a mechanically-driven water pump with an electric one, with the additional benefit that it can be regulated and operated independently from the engine, approximately 0.2l/100km of fuel can be saved."*

Materials used in these types of applications tend to be engineering plastics such as nylons (PA) and polybutylene terephthalate (PBT) because they offer either easy processing and assembly (PA6) or dimensional stability and high dielectric strength (PBT).

Under-hood electricals are also beneficiaries of plastics. Noryl and Valox from SABIC (**www.sabic.com**) are the choice for some automotive parts suppliers for use in power train solutions to displace weight and increase performance, particularly in plug-in hybrid and battery electric vehicles. Although both of these types of vehicles can reduce fuel consumption and emissions, the additional weight of battery packs – up to 300kg or 660 lbs. on a mid-sized car – can undermine environmental benefits. To reduce weight, plastics can replace steel in different applications such as battery housings, which are already being adopted by auto manufacturers. Noryl and Valox also offer chemical and temperature resistance, dimensional stability and flame retardance.

Automotive will always offer good opportunities for custom molders who understand that industry's requirements and have the financial flexibility to play by that industry's rules with respect to quality requirements and payment terms.

> ☞ **For years, the value of electric and electronics content in automotive has been growing between 5-10% per annum.**

7.10.2 Computer and connector applications

Solid State Disk Drives (SSDs) have experienced strong market growth as they contribute to higher data reliability, lower weight and noise, faster access times and power reduction. Serial Advanced Technology Attachment (SATA) and Serial

Attached SCSI (SAS) are set to be the dominant interfaces to interconnect SSDs to notebooks or the strong growing netbooks.

Molex Inc. (**www.molex.com**), headquartered in Lisle, IL, a leading supplier of interconnect products, was seeking a halogen-free insulation material. They selected the F11 grade of Stanyl ForTii polyimide from DSM for their latest high speed SATA connector, because it offered the highest pin retention forces, as well as a range of key processing and performance benefits.

DSM have expanded production capacity for the material, noting that Stanyl ForTii meets market trends for miniaturization and convergence of electronic devices like cell phones and computers. It is suitable for connectors and sockets in these applications because of its dimensional stability and compatibility in lead free surface mount assembly processes.

Molex also use Ticona's Celanex XFR 4840 halogen-free PBT for connectors. These particular components were previously made from another PBT material but the change fulfilled regulatory requirements to eliminate certain brominated flame retardants.[15,16]

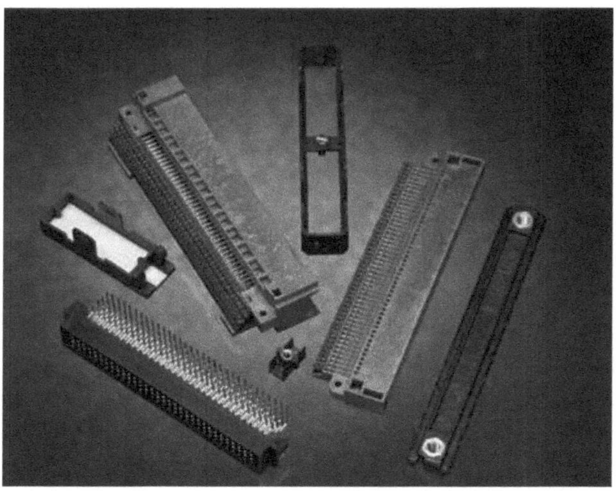

Figure 7.7

These multi-pin connectors require precision moldmaking and molding to get the necessary precise, tight-tolerance pin placement *(RTP Co.)*

15 Restriction of the use of certain Hazardous Substances, RoHS Regulation (Directive 2002/95/EC).

16 Directive on Waste Electrical and Electronic Equipment, WEEE (Directive 2002/96/EC).

7.11 Consumer and personal care products

While consumer spending has dropped in recent years, personal care items have generally held up reasonably well. It seems that personal care is one of those areas in which consumers are willing to spend even in a down economy. (Personal care products are typically low value and include razors, cosmetics, toothbrushes, hair care products such as combs, brushes, and other items that people use every day in personal grooming.) However, many consumer companies still cut back on production and cut prices to lure customers to buy, while at the same time developing new products, new designs and new ads to push consumers into spending.

The big personal care OEMs like Unilever and P&G have extensive supply chains that incorporate mold manufacturers and custom molders, and require a breadth and depth of capabilities. For moldmakers, that generally means expertise in prototyping and R&D assistance, multi-cavity molds that include stack molds, and rotating stack molds, two-shot (two-material) molds and hot runner molds. For custom molders it means having the capabilities to run these molds, including two-shot and two-material molds for over-molding applications, as well as having additional secondary operations such as hot-stamping, embossing and other value add services.

7.12 Summary

There are many choices on offer when deciding on which markets and industries or companies within those markets or industries you are best suited to serve, given your capabilities, your expertise, and your size. You can start to research the markets you think you want to serve by looking them up online. There are many resources that provide in-depth market research information including the number of companies, the biggest players, where they are located and what they require. You can also research by type of product such as medical devices or plastic containers, food packaging, etc. **Study the companies in the markets you choose, and the locations and demographics of their supply chain and the customers they serve.** You may also find it worthwhile to purchase research reports on these markets from reputable market and industry research firms, so that you know who the players are in the markets you choose, and also where commentators see their business moving in the coming years.

As discussed previously, you are looking to **match your strengths with potential customers' weaknesses, or those areas where they plan to grow their business.** Once you have your strategic marketing plan in place and have researched the companies, you'll be able to know which ones are a good fit for your capabilities and expertise, and can maximize your resources. This type of strategic planning is more important today than ever, given the global competition and the number of custom molders all vying for the same business. If you feel that you do not have the resources within your company to do this, you can always hire someone to do this or an outside firm who can gather the information for you.

It's critical to the success of your business – now more than ever – to **create a niche for yourself** with the markets you want to serve, and then to serve that niche better than anyone. That is the way to develop the relationships required to have long-term, successful partnerships with your customers, and to become known in specific markets among companies in those markets, as a supplier of first resort.

Chapter 8

Business challenges for the 21st century – local or global?

The dichotomy for many molders is and always has been the fact that they are small family-owned businesses (in comparison to the large corporate OEMS), that are typically long on molding expertise and experience, but short on business savvy. Understanding how to deal with the large, multi-national corporations is difficult for many of these suppliers, and has ultimately been the undoing of many of them. It's like David trying to deal with Goliath on Goliath's terms!

Molders typically have had to play by the rules of their OEM customers, many being reluctant to establish their own rules such as payment terms, and unable to come to a meeting of the minds. In recent years price issues have taken center stage, as the global competitive landscape came increasingly to the forefront. With China added into the mix as a major competitor, alongside the arrival of the internet and the resulting revolution in global communications, the entire world quickly became the competition for US and European molders.

The question is how does one play on this global stage? What strategies does one need to be competitive not only with respect to price – anyone can play the price game – but in finding a niche, providing value in that niche, and making the best decisions you can for your company.

8.1 Finding your niche

It was once said to me, and I've repeated this phrase often, that 'niches are riches.' Of course the person who said this to me was very successful in his niche of making pultruded fiberglass composite replacement handles for shovels, rakes, hoes, hammers and garden shears – and almost any implement with a wooden handle that eventually wears out and breaks.

Sometimes niches develop as a result of getting work in one type of part for many companies, or from molding many parts for one company or one market. Niches can also develop around material types, such as molders that only mold tough engineering resins such as nylon, polyamideimide, polyetherimide, and other difficult materials.

☞ **Niches are riches.**

Case history: ABA-PGT Inc.

Many times, developing a niche and expertise in one type of component can lead to new business molding the same general types of components for other companies in other markets. For example, ABA-PGT Inc. was a fully integrated producer of precision tooling and high-volume tight-tolerance plastic gears (**www.abapgt.com**). The company built its moldmaking and molding business around gears. Gradually, the company found the same type of precision, technical parts in other industries including the automotive, water management, HVAC and business machine/paper path market segments.

Another industry that requires small, precision parts is the medical device industry. To help increase business in that market segment, ABA-PGT obtained its ISO 13485:2003 Medical Device Certificate. While not solely dedicated to the medical device industry, Terry R. Holmes, vice president of Sales and Marketing, said the Medical Device Certification adds another level of credibility and value to ABA-PGT's capabilities. *"The majority of our business is to produce the tooling and provide the molding technology for plastic gears, which is our core area of expertise. For several years however, we've been doing non-gear precision parts in the medical device market that often lend themselves to the expertise we've developed in small, precision gear parts. That's how we entered into the medical device field – through our high level of precision."* The company operates with 52 molding presses ranging in size from 17 to 110 tons, and employs approximately 100 staff. Holmes said that the company is looking to expand its tonnage range going forward. However, the size of the presses has also helped the company develop a niche that is unique.

A niche can also be carved out through certifications. Obtaining its ISO 13485:2003 Medical Device certification gives the company a competitive edge when approaching companies in the medical device industry, Holmes explained. *"I think the greatest benefit to having the medical certification is so that additional medical device companies will feel more comfortable working with us."*

Every molding company has a niche, but sometimes it takes doing your homework – your strategic marketing and business planning – to uncover just what that niche is. Then it takes more planning to capitalize on the expertise you've developed in that niche to locate other business looking for what you can offer.

Case history: BNL (UK) Ltd.

Sometimes molders intentionally develop a technological niche by invention like BNL (UK) Ltd., which in 2010 celebrated 40 years as a supplier of plastic bearings (**www.bnl-bearings.com**). The company, with headquarters in North Yorkshire, UK, was founded by David Garnett, who is widely acknowledged as the inventor of the molded polymer bearing, or 'bearing no-lube' – hence BNL.

BNL's initial product range of lubricant-free, thermoplastic rolling element bearings were key components for products such as conveyors, food processing equipment and furniture, to name a few. However, the injection molding process allowed BNL to quickly move from producing standard radial bearings to integrating molded bearing tracks, and into pulleys, skate wheels and end caps. The company subsequently moved into more complex molded bearing designs and assemblies for applications such as photocopiers, security cameras and pool cleaners.

BNL has a long history, including being sold to Henry Barratt plc in 1988, and Garnett returning as Managing Director, then repurchasing the company in 1991. In 1995, BNL because part of the Swiss Sarna Group until current parent company, Plastics Capital, acquired BNL in 2005, and today has plants in Foxboro, Massachusetts (USA); Tokyo, Japan; and Rayong, Thailand. Throughout all these changes of ownership, BNL has remained focused on its unique product lines. Reflecting on the company's development, Neil Partlett, CEO of BNL in 2010, has commented, *"During a long and illustrious history, BNL has been a pioneer in its field and enjoys a worldwide reputation as a provider of innovative plastic bearing solutions."* Today BNL continues to design and manufacture innovative thermoplastic bearing assemblies and provide technical molding solutions.

BNL is a good example of a molding company that began with a niche through an invention, stayed true to its niche and developed new business as advancements in mold and molding technology allowed it to move forward into new products for a variety of customers in diverse industries. Not all molders have a proprietary product, but both ABA-PGT and BNL are good examples of how molders can find or develop their niche.

8.1.1 Proprietary product niches for custom molders

Custom molders realize that machine utilization depends upon how busy their customers are, which is beyond their control. To help level the ups and downs of the custom molding business, some molders have sought out a line of proprietary products in order to keep the presses busy. **In some cases this is a tough road, because while these companies know a lot about moldmaking and molding, they may know little about the business that goes along with owning a product line, including marketing, sales, packaging and distribution**.

One company that has been successful is Jones Plastic and Engineering Co. LLC (**www.jonesplastic.com**), which recently opened a 178,000-square-foot dedicated injection molding operation at its Jeffersontown, KY, facility to produce the company's Rev-A-Shelf line. The company installed eight injection molding machines, moved from elsewhere in the business to accommodate the molding.

Jones established the Rev-A-Shelf division in 1978, and it proved to be a fairly recession-proof business over the years. In mid-2009, Rev-A-Shelf acquired more proprietary household lines from Peter Meier Inc. of Kernersville, NC, and Jones brought that molding work in-house from custom molders.

Finding a proprietary product niche if you're a custom molder isn't easy, and it's even more difficult to juggle the proprietary work with the custom work. Some proprietary product molding operations even end up as custom molding businesses, because just as it is attractive for custom molders to make a proprietary product to level out the ups and downs of the custom business, so equally some proprietary product molders find that when demand is down for their product, they have open press capacity they need or want to fill. Thus, they take on custom work to boost their revenue. It's actually a fine line and just finding a proprietary product line isn't the answer to leveling your business.

8.2 Competing in a global market

The new millennium brought increasing competition from every corner of the world. Not that there wasn't global competition prior to that – there certainly was! However, it hadn't really hit the world of smaller companies such as custom molders and mold manufacturers. Global competition had largely been reserved for major OEMs whose products were sold around the world, and who were putting facilities in those countries.

In the 21st century competition from Asian countries really began to affect these smaller companies, both in molding and moldmaking, with the sheer scale of manufacturing in China ensuring that its impact was the greatest. Mold manufacturing was generally hit hard, both in North America and Europe, by Chinese mold manufacturers whose prices could be as low as 30% of North American prices. Mold buyers from the large OEMs tend to buy on price, and that was particularly true if the mold was to stay in China for the molding. In addition it was widely felt that if much of the mold manufacturing went to China, then it wouldn't be long before the molding would follow. This led some larger custom molders to form joint venture partnerships with Chinese molders and others to establish plants there.

These companies typically moved in to serve the OEMs, which had seen the potential of a huge and increasingly affluent customer base. Some of these ventures worked well, and some didn't. It was often more difficult for the smaller molding company, and most needed to hire management that knew how to deal with the various governments in the provinces where plants were located. They generally also discovered that language and cultural barriers were quite high.

8.2.1 Locating global facilities

According to George W. Blank, CEO and Group President of MedTech Group LLC (**www.ventionmedical.com**). *"You want to be reasonably close to your market, primarily because there's a lot of 'non-value' added in shipping parts. Some injection molders would tend to follow customers and that makes a lot of sense."*

Moldmaker, molder, and medical contract manufacturer MedTech Group, headquartered in South Plainfield, NJ, is a company that specializes exclusively in the medical/surgical, pharmaceutical, diagnostic, and biotech sectors. In February 2010, the company announced plans to establish a 'best-in-class' moldmaking operation at its existing molding facility in Heredia, Costa Rica. Its New Jersey mold-building resource was to focus on tooling design and engineering, as well as overall technology advancement.

Blank said that it seemed natural to bring a full range of tool-building capabilities into Costa Rica to service MedTech Group's customers there, adding that this was in line with the company's goal of enhancing value and speed-to-market for all its customers worldwide. *"We have had a very positive experience thus far in Costa Rica, and have found the skill level, dedication and craftsmanship of our employees to be exceptional,"* Blank stated.

MedTech began operating in Costa Rica in 2004, and in 2009 opened a second facility there to manufacture complete medical devices. The company's facilities now total 60,000 square feet and include a clean room molding area, assembly, and packaging services for the medical device industry. MedTech, founded in 1979, is truly a global company today with facilities in West Haven, Connecticut; Vega Baja in Puerto Rico and Singapore.

Traditionally, US molders have been less likely than their European counterparts to set up in new countries, as the sheer scale of the United States offered so much more potential closer to home. As noted earlier, many larger custom injection molders began siting plants in other regions of the US during the boom years of the 1990s. A few were even daring enough to put plants in Mexico – near shoring, as it came to be called. The big moves came about when they finally began to venture 'offshore' to places such as Taiwan, mainland China and Singapore, all in an effort to be near their customers and provide better service.

☞ ***"There's a lot of 'non-value' added in shipping parts."***

In October 2009, medical injection molder Classic Industries Inc. (**www.classicind.com**) headquartered in Latrobe, PA, acquired RW Injection Moulding Ltd., located in Middleton-in-Teesdale in the Northeast of England. The acquisition gave Classic Industries an operation in the United Kingdom to serve medical device and health-care customers throughout Europe. In addition to the company's US headquarters facility, Classic Industries operates plants in El Paso (Texas), Ponce (Puerto Rico), and Shanghai (China).

Learning to play globally has been critical to many of the larger molding companies, and many of those have become global companies over the past two decades. As we have already seen, some molders began their global foray by heading to Mexico – or at least to the Texas/Mexican border to take advantage of the Maquiladora or 'twin plant' trade opportunities. The Maquiladora program allows molders to serve their OEM customers who have manufacturing plants along the border, without actually going into Mexico themselves. However, quite a few molders did eventually locate in Mexico.

North American OEMs typically build large plants in Mexico within 'supplier parks' so they can bring their supply chain with them. Ford Motor Co., for example, has a large supplier park in Hermosillo, Mexico, designed to make the supply chain shorter and easily accessible. This prevents downtime due to shipping problems across the border.

Canadian manufacturer Bombardier Aerospace operates a world-class manufacturing facility in Queretaro, Mexico, which produces electrical harnesses for all Bombardier aircraft currently in production. The facility has 1,100 employees that also produce structural aircraft components. They too have encouraged their supply chain to move onto the campus that was established in Queretaro.

Molders are responding cautiously to the growth of major manufacturing companies in Mexico, but responding nonetheless. Some are locating plants in Mexico in the form of joint ventures, such as Pyramid Plastics Inc. (**www.pyramid-plastics.com**), which formed a joint venture in March 2006 with proprietary injection molder Roi Plastics S de RL MI of Monterrey, Mexico. According to Pyramid's General Manager, the joint venture was a way for the company to serve its customer base in Mexico and position the company for future business opportunities there.

Global injection molder and contract manufacturer Nypro Inc. (**www.nypro.com**), headquartered in Clinton, MA, has three facilities in Mexico, including its latest one, a 97,000-square-foot manufacturing plant in Chihuahua, called Nypro Chihuahua. The facility has capabilities for automated molding and assembly, mold building and repair. The company serves customers in the electronics, telecommunications, consumer, industry and automotive markets in the region. Nypro also has plants in Juarez (across the Rio Grande from El Paso, TX), Guadalajara and Monterrey.

8.2.2 Changes to the rules in Mexico

Companies looking to establish facilities or joint ventures anywhere in the world must obviously take advice about tax laws and other business conditions in that country. Before establishing a plant in Mexico, it is important to be aware of the latest changes to the laws in this area. According to Plante & Moran PLLC (**www.plantemoran.com**), an accounting and business advisory firm with global services, at the time of writing the Mexican Ministry of Economy was proposing certain changes to the Mexico's IMMEX program (the formal name for the maquiladora or 'twin plant' model, i.e. one on each side of the border). If approved by the Mexican Congress, these changes could have a significant impact on some IMMEX operations. In a February 2010 update the company stressed that these changes were still only proposals, and not part of the current rules and regulations in Mexico.[1]

"*The goal of these changes appears to be to limit many of the program's benefits to only those companies that operate like 'traditional' maquiladoras,*" stated Scott Sneckenberger of Plante and Moran. "*This means companies that operate as follows:*

- *The company acts as a toll manufacturer only – their one and only customer is their foreign parent company. No direct sales to third parties occur from the IMMEX operation*.
- *The IMMEX operation is not the owner of the machinery and equipment used in its operation, nor has it ever been the owner of such items.*
- *The majority (more than 50%) of the parts or subassemblies utilized in the IMMEX operation are sourced outside of Mexico*."

For 'traditional' IMMEX operations, the tax landscape in Mexico would not change. However, for companies not meeting such requirements, the tax implications could be significant both for the Mexican operation and its foreign parent.

1 Potential Changes to Mexico's IMMEX (Maquiladora) Program, March 3, **2010**, www.plantemoran.com.

8.2.3 Offshoring and back-shoring

Offshoring has been a thorn in the side of moldmakers and molders for nearly two decades, but never like it's been in the new millennium. The relentless drive by OEMs to reduce manufacturing costs has impelled them to seek 'cheap' tooling and molding sources in China more than any other country. Interestingly, tooling which has come back to the US and Europe from Chinese sources has proved to be a mixed bag - some good tools and some that are little more than boat anchors - and consequently OEMs are learning some valuable lessons about the value and quality of US and European molds and the intellectual property that goes with them.

A 2009 report from Archstone Consulting LLC[2] suggests that while for more than a decade, "*offshoring production and suppliers has been the mantra for cost reduction*," that is changing as a new paradigm comes into play. Offshoring may not be the route to cost reduction that it was once thought to be. According to Archstone, there are 10 macro forces currently undermining the offshoring trend:

1. Uncertain global economic future
2. Volatile commodity prices
3. Continuing high offshore transportation costs
4. Exchange rate uncertainty
5. Increased offshore labor rates
6. Higher cost of quality
7. Reduced supply chain flexibility
8. Decreased visibility and increased network complexity/risk
9. Increasing customer requirements
10. High intellectual property theft and piracy/grey market

Based on an in-depth survey of 39 senior executives from US and European manufacturers, Archstone reported that nearly 90% of manufacturers said that they were considering or had already begun rebalancing their manufacturing

---

2 Ferreira, J. and Prokopets, L., Does Offshoring Still Make Sense?, Supply Chain Management Review, Jan./Feb. **2009**, pp 20-27.

and supply strategy; 40% reported an increase in direct costs of 25% or more in the previous 3 years; and 90% expected further increases of 10% or more in the coming year.

Many of these factors are unpredictable, and entirely beyond the control of the molder. The point about supply chain flexibility is an important one in some market areas. Manufacturing overseas tends to mean that products must be shipped in container loads, reducing the molder's ability to respond quickly and flexibly to changing customer needs. Other problems concerning product quality and intellectual property have been widely reported, particularly in relation to China.

Moldmakers and product OEMs especially, have to be careful about outsourcing molds for specific products, particularly consumer products that have a 'mass market' appeal. In some cases, the Chinese mold manufacturer has reportedly built one mold for its US customer and another mold for itself. In a case reported in the US press, one entrepreneur and inventor of the 'BraBaby'– a plastic ball with many openings in which to put bras for washing in automatic washers (**www.brababy.com**), claims that she has been fighting off counterfeit 'BraBaby' products for years after having her mold made in China and shipped to the US for molding.[3] Intellectual property theft continues to be a problem with offshoring, and many OEMs find that it's difficult – if not impossible – to enforce patents, copyrights and other laws that are meant to protect products from counterfeiters. Even industrial products such as automotive aftermarket parts have been subject to counterfeiting.

Case history: Diagnostic Devices

Medical products manufactured in China have also reportedly proved to be problematic from a quality standpoint, which can result in lawsuits over products that fail to meet FDA good manufacturing practice (GMP) standards. Controlling a medical manufacturing operation in China from a US headquarters is not easy. Diagnostic Devices Inc. (**www.prodigymeter.com**) manufactures blood glucose monitoring systems called Prodigy, and at one time had 600 workers at its China facility. However, problems with quality control, rising costs and concerns over its intellectual property prompted Rick Admani Abulhaj, COO, to bring manufacturing back to the US. "*Using cutting-edge automation and robotic technology will give us greater control of our operations, reduce costs and protect our intellectual property. The primary benefit to taking manufacturing to China is for the labor, not for the technology. Robots, vision systems, other automation – even if you put all that in China there's no one there who can fix the machines or service them. You go for cheap labor.*"

3 Cheng J., A Small Firm Takes on Chinese Pirates, Wall Street Journal Online, July 5, **2007**, www.wsj.com.

However Abulhaj also noted that throughout his career he's been instrumental in taking products from the US to China. "*We once had 4,000 employees and did all types of manufacturing including injection molding, extrusion and assembly for high volume products all related to the health-care industry. What drove us offshore was the startup of a new product – it is much cheaper offshore because the volumes are low. But if you have high volume products, it makes more sense to manufacture them domestically especially if you have automation. If we can automate the product here in the US, it's foolish to take it to China. We can do it at home in our back yard for the same cost, and we have control over our intellectual property (IP).*"

Quality control is one piece of the offshoring puzzle that no one talks about, notes Abulhaj. In his experience "*You can specify a certain PMS (Pantone) color, but they run out of that color so they mix their own and give you something that looks close to the color you specified. Run out of the material you specify so they use another material that they believe is close to the material you need,*" he said. "*We have to adhere to FDA regulations. When you make products in the US, you make them to a higher standard, particularly in health-care the FDA is the law. If you don't comply – if you do things wrong here – you get your products recalled. In China there are no ramifications.*"

Regardless of how good manufacturing practices in China become, Abulhaj claims that the major benefit for Diagnostic Devices' products is the assurance of quality of a product made in the USA. "*Everyone, even people offshore, understands the higher quality standards of US-made products. A US product is of a much higher quality, and the consumer is driving us toward more US products.*" That said, Abulhaj also added that he couldn't afford to make a low-tech product in the United States.

However, with labor costs in China escalating at the time of writing and new regulations being implemented, together with quality issues, shipping, the difficulty of managing an extremely long supply chain, and other challenges for firms operating overseas, **the advantages of outsourcing appear to be shrinking**. US moldmakers now say that bidding against Chinese mold manufacturers is not necessarily a losing proposition, and that the price differential between US and Chinese molds has narrowed from 50-60% to 20-30%. Savings that small can easily be eroded by shipping costs, and other costs associated with getting a mold from China.

Even as early as 2008, according to a report in The McKinsey Quarterly (September 2008) companies were already beginning to rethink their offshoring strategies. Oil spikes in recent years sharply increased the price of raw materials and shipping. The diminishing wage differentials between Chinese and Mexican workers and those in the US and Europe, also served to narrow the gap. The article quotes the example of Mexican workers earning twice as much as their Chinese competitors in 2003, which resulted in many companies shifting work from Mexico to China. By 2008 the balance had shifted significantly, with Mexican earnings just 1.14 times as much.[4]

AlixPartners LLP, a global business advisory firm from Southfield, Michigan, compiles and publishes a regular Manufacturing-Outsourcing Cost Index, intended to show manufacturing company bosses not just which manufacturing locations are low-cost countries (LCC), but also best-cost countries (BCC). Looking at a basket of components and assemblies (not just plastic items) the index compares the cost of building these items in the US and various competitor countries. The survey also tracks changes in seven key cost drivers. In 2008 the position of China relative to the US was declining, with total parts cost in China reaching 94% of US costs. However in 2009 the Chinese position strengthened versus the US. Mexico and India led the rankings in both years as the most cost-effective manufacturing locations. Interestingly, Vietnam, Russia and Romania were added to the survey in 2009 and all outperformed China.

Comparison of lowest-cost countries for US manufacturing outsourcing

Ranking	End 2008	End 2009
1	Mexico	Mexico
2	India	India
3	China	Vietnam
4	US	Russia
5	Brazil	Romania

AlixPartners US Manufacturing-Outsourcing Index, Feb 2010, *http://www.alixpartners.com/en/MediaCenter/IntellectualCapital/tabid/89/language/en-US/Default.aspx.*

This kind of detailed study flags up the variability between locations in some less obvious cost areas, such as energy charges, taxation and construction costs, as well as the more obvious material, labor and shipping costs, and exchange rate fluctuations. It emphasizes the volatility of the situation, and the need for monitoring and flexibility if companies wish to benefit from global variations rather than fall victim to them.

While the US dropped in competitiveness in 2009, along with a number of European countries, individual companies in molding and moldmaking can be competitive and help their OEM customers to be competitive as well, through new molding and machine tool technology that increases productivity and efficiency, and reduces labor and energy costs. There will always be competition, LCCs will

4 Goel, A. and Moussavi, N., Time to rethink offshoring?, McKinsey Quarterly, September, **2008**, www.mckinseyquarterly.com.

shift as the economy shifts, and BCCs will be those that encourage innovation, entrepreneurship and flexibility.

Figure 8.1

Farouk Systems, manufacturers of professional hair-care products such as the patented CHI hair iron, brought its products back to the U.S. from China due huge intellectual property theft problems. It was costing the company upwards of $500,000 a month to battle the counterfeiters, including replacing defective CHI hair irons and dryers that turned out not to be manufactured by Farouk Systems

Chapter 9

Business challenges for the 21st century – sustainability

Sustainability is a word that we hear often in connection with almost everything. In fact, sustainability has several connotations, and for many processors confusion reigns when it comes to understanding what it means to be a 'sustainable' manufacturer. What's the hype? What's the reality? Many molders are greener than they think they are!

In fact, most molders probably don't think of themselves as 'sustainable' manufacturers, but whether they recognize it or not, every company is engaged in some activity that qualifies as sustainable That's the opinion of Larry Buck, president of IML Solutions LLC, (www.iml-solutions.com), speaking at a recent IMLCON conference on in-mold labeling and decorating. Buck noted the recycling trends and the large amount of plastics that is being recycled, but added, *"We have to develop more recycling streams to handle the amount of materials available for recycling."*

Buck said that the two major goals for plastics companies include **reducing material usage** through light-weighting, thin-walling, processing technologies such as MuCell, and additives for strength; and reusing products. *"Many products can be used for other than what they were originally intended for,"* Buck said.

Additionally, there is recycled content to consider, but with some caveats. Post-consumer recycled materials have to be sorted and many cannot withstand commingling. Some technologies make recycling more complex, such as in-mold labeling or decorating. *"You can use post-consumer recycled content but if you mix your products, you complicate the recycling stream,"* Buck told attendees at IMLCON. *"Your label must be the same as the package material."*

> ☞ **Every company is engaged in some activity that qualifies as sustainable.**

Molders are becoming more aware of most of those goals, and customers are working with them to drive out costs or enhance recyclability through a variety of processing technologies such as co-injection. Cascade Engineering (www.cascadeng.com) launched its Eco-Cart in mid-2007 through the company's Container Div. The Eco-Cart is a durable solid waste recycling container made with 30%-50% post consumer recyclate (PCR) content. Using the co-injection process, Cascade was able to solve the problem of incorporating a high amount of recycled content into the containers, yet making them rugged and durable enough to withstand weekly pickups by automated garbage trucks.[1]

Many molders engage in recycling efforts for everything from scrap paper to water and soft drink bottles. Jatco Inc. (www.jatcoinc.com) for example, began a company-wide recycling program nearly two years ago that involved all of this custom injection molder's 200 employees. Funds garnered from selling everything from recycled office paper, shrink wrap, straps from gaylords, corrugated materials and more, were returned to the employees. Jatco's goal: to become a zero landfill company.[2]

For some molders, **sustainable manufacturing can be as simple as changing out lighting throughout the plant, and installing skylights over areas such the production floor and the warehouse to use more natural light.** Buying newer, more energy efficient equipment to reduce electricity usage is also popular.

Molding companies want to manufacture in a sustainable way for many reasons. Not only does it help them reduce overall costs to manufacture and make them more competitive in the global marketplace, their efforts can also make them attractive to OEMs who want to use 'green' suppliers to help in their own efforts to reduce their carbon footprint. With some big-box retailers such as Wal-Mart using score cards to track suppliers' sustainability, more OEMs are actively seeking out suppliers that can support these efforts.

Another advantage to the efforts being made by US OEMs and molders is the 'Made in the USA' label. The entire supply chain – from OEMs to retailers to consumers – is becoming conscious of the efforts being made in **reducing the carbon footprint of products**, resulting in some OEMs bringing manufacturing back to the US from China for example.

Nypro Inc. (www.nypro.com), a global contract manufacturer of custom plastic components with 1500 presses at 49 locations in 16 countries, began nearly two decades ago putting manufacturing plants wherever their customers needed their suppliers. Richard Moore, Global Director of Technology, commented several years ago, "*We are seeing a shift now, from everything going toward Asia to wanting to manufacture more products locally for the large US consumer base. Not that it's all coming back from China, but with respect to the issue of sustainability, there is*

---

1 Goldsberry, C., Cascade's evolution includes new partnership, sustainable products, October 8, **2009**, www.plasticstoday.com.

2 Goldsberry, C., Greening moves from option to opportunity, December 8, **2009**, www.plasticstoday.com.

a lot of concern by the big brand name companies with respect to carbon footprint, and consumer concerns are driving this."

The key to the effectiveness of sustainability in manufacturing is after all, the bottom line. As Cascade's Samia Brown pointed out, *"Fred Keller [Cascade's CEO] has always believed in sustainability, and while business should always do the right thing, it has to make good sense from a business point of view."* Jatco's Steven Jones would agree and commented, *"You have to do the right thing for the right reason, and when we did that we were rewarded with benefits we didn't anticipate."*

Molders are engaged in sustainable activities whether or not they realize it. The number one focus of sustainability in manufacturing operations, according to a November 2009 report from the Aberdeen Group is **energy efficiency**.[3] Ninety-four percent of the 177 respondents to Aberdeen's survey, responded that energy efficiency is their top focus. "*One of the reasons energy is on the top of the mind is because it directly impacts the company's bottom line,*" according to the report's author. "*This is especially critical in an energy-intense plant where energy cost is a large percentage, often upwards of 25% of the total operation costs of the plant.*"

Molders understand this cost, as molding machines and other auxiliary equipment draw a significant amount of energy. Many are choosing newer, more energy efficient equipment to realize greater energy savings, or retrofitting older equipment with energy-saving devices. However, there is still much more that can be done here, for example looking for areas where energy is wasted through poor plant design and maintenance, inappropriate machine and temperature settings, machinery and ancillaries left running when not required, and many other common practices!

Waste reduction and environmental impact were the next most selected focuses for manufacturers among 78% of respondents in each category. Reuse and recycling came in third cited by 72% of respondents as a focus at their manufacturing facilities. Employee safety came in as a focus with 70% of the respondents, but the Aberdeen survey showed that companies are also "*expanding their sustainability initiatives to holistically manage the safety of employee, process and product.*"

Aberdeen reported that one area that has been growing in importance over the past few years is that of governance, compliance and reporting. "*The emergence and evolution of regulatory mandates in a growing number of geographies related to the trade, reporting, and/or disclosure of emissions has compelled companies to make governance, compliance and reporting an integral part of their sustainability strategy,*" said the report.

3 Shah, M., Sustainable Production: Good for the Plant, Good for the Planet, September 30, **2009**, www.aberdeen.com.

Over the years, Aberdeen has tracked a trend among manufacturers in which sustainability has changed from being a management philosophy that is 'nice to have' to one that is core to the success of the organization. Proof of that trend came when in the most recent survey, only 10% of respondents said their companies are focusing on sustainability because their competitors are. This 'me too, because I have to' mentality is fading, concluded the Aberdeen report.

9.1 Lean manufacturing and sustainability

Case history: McClarin Plastics [4]

McClarin Plastics (www.mcclarinplastics.com) has found a way to merge green ideals with lean principles, and found that it can have an impact far beyond raw materials, water and renewable energy. ('Lean' describes the ability to achieve more with fewer resources, by the continuous elimination of waste in every area of the company, from the front door to the shipping floor. This widely adopted principle originated in the Toyota Company in Japan.) Thinking outside the 'green' box, McClarin found that forging sustainable relationships within a supply chain, along with the 'conservation' of established partnerships, can produce a positive impact on the bottom line as well. McClarin, a 50-year-old company, is a leader in thermoforming and vacuum infusion of fiber reinforced plastics, coordinated a cooperative lean certification session for their employees, customers and suppliers.

According to a study commissioned by the US Environmental Protection Agency (EPA), suppliers in lean supply chains which deliver a component in the right quantity at the right time, share the benefits of reduced costs as well as waste reduction, and can deliver higher quality parts. To achieve the true value of 'lean', a company needs to involve the entire supply chain, otherwise it is estimated that an individual company can only lean their operation by 25-30%.[5]

"*Each segment of the supply chain must understand the others' needs. One kink in the chain can throw off the entire process causing waste and expense as well as excessive use of energy and raw materials,*" said Roger Kipp, vice president of marketing and engineering for McClarin Plastics. "*This will bring everyone involved in a related supply chain together to learn how their performance affects others. The positive bottom line impact from the resulting relationships and understanding could be huge.*"

---

4 www.mcclarinplastics.com/files/pr 04.pdf.

5 Womack, J.P., Jones, D.T. and Roos, D, The Machine That Changed the World, **2007**, Simon & Schuster Ltd, 978-1847370556 .

Since McClarin formally adopted the lean manufacturing methodology in 2000, the company has continually realized green benefits in the form of space allocation, waste reduction, energy conservation and increased cash flow. Lean facility management produced some of the most impressive results. The company has tripled its sales volume since the implementation of lean, while only increasing its square-footage by about 25%. This is due in part to the Lean Continuous Flow Work Cell concept, in which the complete production of a component occurs in one location. This has resulted in 30,000 square feet of warehouse space being freed and reallocated as production space because raw materials are delivered just-in-time (JIT) directly to the work cell.

The US EPA study also found that consistent product quality is a basic aspect of competitiveness, which affects cost and customer loyalty. When all members of a supply chain are operating under the same guidelines, i.e. lean, the end product presented to the purchaser is of a higher quality with timelier delivery and in some instances, lower costs. When suppliers are not operating as a team, product defects escalate and costs compound due to added time, labor and space for rework and repair, material waste and disposal. This impacts both the environmental and fiscal bottom line, as recurring defects and delivery delays can mean the loss of lucrative contracts and a more substantial eco-footprint.

McClarin notes that it has seen increased product quality and expects to see even more improvement once their entire supply chain is practicing lean. This translates to reduced waste from defects, overspray and scrap being sent to landfills. However, McClarin recycles about 95% of the materials that the company deems recyclable. Energy is another area where lean has produced green benefits. McClarin has leveled production activity to customer demand and has in turn lowered spikes in energy demand. Additionally, the company manages machinery starts and stops around non-energy spike times.

"We are anxious to share what we've learned about eco-responsibility through using lean principles," said Kipp. *"Our hope is to get everyone in the supply chain operating on the same page so they too, can realize the benefits. We've lowered overhead and increased cash flow, which we've reinvested back into the company."*

McClarin's cooperative lean certification session for its supply chain was the first in the state and had multiple goals, including contributing to eco-friendly initiatives by reducing raw material consumption, energy and inventory. However, McClarin noted that the main goal was to encourage the use of lean principles by all segments of a supply chain to reduce waste, human effort, manufacturing space and time, which in turn will reduce supplier turnover and the costs – fiscal and human – associated with locating and training new suppliers.

9.2 Energy and sustainability

For molders worldwide the cost of energy is a huge issue because molding machines use a lot of energy. While some companies have turned from hydraulic to all-electric machines for greater energy efficiency, running presses still isn't an inexpensive proposition. Many presses are equipped with servo motors that help reduce demand between cycles, and that can also contribute to overall energy efficiency. However there are things to consider when trying to conserve energy.

For example, machines pull more power on startup, so many molders try to run 24/7 in order to keep the power load steady and reduce spikes in energy pulls. That might mean running shorter-run jobs longer and storing inventory, and making releases from inventory as the customer demands instead of running just what is needed each month or each quarter. (Arguably though, inventory costs money, increasing working capital, damaging cash flow, and requiring storage space and staff to manage it. In the worst case scenario, you could also be left with unsold product if you lose the business or customer. As in many other areas, this is a balancing act requiring careful analysis.)

Some companies have put skylights into their roofs so that more natural light can be used during the daylight hours, conserving the lighting for second and third shifts. There are also new energy efficient lights available that are easier on the eyes, and make the production floor brighter while saving on electricity. Installing motion detectors in the offices, storage rooms, restrooms, and other areas that are not used 100% of the time also helps in turning lights off when people leave the area. While this sounds like a little, it can end up saving a lot when it comes to the electric bill.

The subject of energy management (and subsequent reduction of energy use) in the plastics industry is comprehensively covered in another Plastics Information Direct publication written by industry expert Robin Kent.[6] Dr Kent has analysed where energy is consumed within a plastics processing operation (based on real data and observations), and offers many practical solutions which can typically produce overall cost savings of up to 30%. **Whilst the actual operation of plastics processing machinery is by far the largest energy user, he points out that ancillary services such as compressed air and chillers are also major consumers.**

---

6 Kent, R.J., Energy Management in Plastics Processing: strategies, targets, techniques and tools, **2008**, Plastics Information Direct, 978-1906479039.

Case history: Mack Molding

Mack Molding (www.mack.com) a leading custom molder for the medical, commercial, business equipment, and transportation industries, has installed more than 2,100 energy-efficient lighting fixtures in its southern Vermon manufacturing facilities. Mack Molding implemented the lighting upgrades in three facilities totaling more than 600,000 square feet. Goals for the project included improving the level and quality of light, saving energy, minimizing maintenance, and achieving a return on investment in two years or less. The company engaged a supplier of energy-efficient lighting for manufacturing environments to conduct a lighting audit and help it determine the potential energy savings.

Because of the age of two of the Mack plants, Cavendish (built in 1890) and East Arlington (built in 1910), new energy-efficient fluorescent technology was installed or retrofitted in every lighting fixture to standardize light levels and to save on energy and maintenance costs. Mack Molding's headquarters facility, built during the 1990s, has ceilings that are about 30 feet above the manufacturing floor. When new energy-efficient fixtures (some with motion sensors) were installed, the results were immediate. Light levels in the warehouse improve by 175%. With the new high-intensity fluorescent (HIF) technology, light levels throughout the plant were increased and hot spots and shadows eliminated.

The estimated electricity savings resulting from the new lighting amount to more than 1.5 million kilowatts per year, or the equivalent of 967 tons of CO_2, or taking 167 cars off the road. An additional by-product of the lighting improvements is the improved comfort and safety of the employees at Mack Molding. It also has resulted in lower operating temperatures, which means less cooling and ventilation required during the summer months. The total cost of the project was US$450,000, but the company expects to recoup its investment within two years.

9.3 Other sustainable ideas

Packaging can also be a large source of waste and additional expenditure if it's not watched. Many molding companies have started using reusable containers, some of which fold flat for return. Returnable packaging is being used by Xten Industries in Kenosha, Wisconsin, with several of its customers, and this brought them a saving of approximately $90,000 from the program in 2009. Some of the shipping containers are large plastic totes which the customer buys. In other cases Xten's customers return the corrugated boxes that are shipped to them a number of times. The program is turning out to be economical for Xten as well. The company has several suppliers – particularly metal suppliers – that now ship their parts to Xten in returnable totes which saved the company over $50,000 in a year.

One of Xten's largest customers asked them to consolidate shipments by delaying delivery until they can fill a truck, rather than shipping small amounts more often, which saved the company about 600 shipments in 2009, equivalent to freight savings of about $120,000 a year. Because Xten molds a lot of very large parts, the company is very aware of the shipping costs involved.

Xten's focus is on sustainability, energy savings, and eliminating waste in nearly every area of its operations. One of the key issues in this area was material usage, which is a common issue for molders, leading Xten to focus on that area of the operations, looking at scrap and **regrind**. One of the goals of the company was to reduce the amount of regrind sold. Rather than grinding runners and part scrap on the production floor, which caused a lot of contamination, dust and noise, Xten created a separate grinding area so that the scrap was cleaner and resulted in more usable material.

Educating customers about using regrind in their products is part of the company's material savings. The company's material planner/scheduler identifies the amount of regrind that can be used on each job. That is put onto a tracking sheet on the floor so that the material handlers know how much regrind can be added. Some customers request virgin material for their components, when actually virgin might not be necessary. That's when Xten offers to provide assistance to its customers by validating parts using a percentage of regrind. "*We definitely work with our customers about the use of regrind. If Xten has a certain family of parts that can use regrind, the company provides the customer with samples to get their buy-in on the savings that they can realize. Customers will not go through these material changes unless there's something in it for them too, and sometimes that just takes educating customers in the benefits of using regrind in parts that may not need 100% virgin material.*"

Another saving in material can come in the **purging** process when changing molds or material colors. Especially where large parts are concerned, a lot of material can be wasted in the purging process. A good purge compound is generally the answer, and that sometimes requires experimenting with the various brands to see which works best for you.

There are many ways for a molding company to be green and sustainability can be good for everyone. In addition to lowering energy costs, reducing scrap of all types, and improving profits, sustainability fits with a company's lean efforts and its Six Sigma program. It provides benefits across the operating platform from the electric bill to the material costs to employee health and safety, creating an environment that is profitable from the top to the bottom line.

9.4 Summary: areas to look for sustainability in your company

- ✓ Reducing materials usage: e.g. light-weighting, using additives to improve strength and reduce part thickness.
- ✓ Recycling: designing molded parts for recyclability, incorporating recycled materials through processes such as co-injection, and recycling other 'waste' materials around the plant.
- ✓ Energy efficiency in manufacturing, ancillaries, factory services and elsewhere .
- ✓ Reducing your carbon footprint, e.g. by minimising transportation of raw materials and end products.
- ✓ Reducing and reusing transit packaging, and rationalizing and consolidating shipping.
- ✓ Managing the use of regrind and the efficiency of purging.

All these actions will potentially improve the 'green' performance and credentials of your company, and also help your customers to achieve their sustainability goals. Equally importantly, when properly managed, they should also contribute to the cost efficiency of your business and help to ensure its ongoing survival and success.

Chapter 10

Business operations for profitability

Many injection molding companies have for years, operated as 'mom and pop' family-owned businesses, primarily because these businesses tended to be small, entrepreneurial companies founded by individuals who were deeply interested in plastics as a new material. Many molding companies actually began as moldmaking businesses that evolved into molding because of the on-going business that could be had from molding the parts as opposed to just building the mold and shipping it off to a molding facility.

Today, economy of scale matters in the molding business. While there are still many smaller molding companies (turning over just a few million dollars or euros), it has become more and more difficult for them to be profitable and stay in business. Indeed, many smaller companies have gone out of business during the first decade of this century. The reasons varied but primarily they did not have the financial wherewithal to withstand the economic downturn, because they either relied too much on one or two customers for their business (you don't need a sales force for one or two long-term customers), or because they had no defined niche – they basically tried to be all things to all people – any part, any price – and therefore had minimal value to those companies looking for value-add suppliers.

That said, many large molders have also fallen by the wayside during this first decade of the 21st century. Some of these were deeply entrenched in the automotive industry as tier two molders, and lacked any diversification that could support them when the automotive industry declined. **The first rule of a successful business model is diversification. That can mean diversifying in markets or diversifying within a particular market with many customers.**

Operating profitably in an industry like injection molding where competition is high, raw material prices change rapidly (and mostly in the upward direction), and with customers that constantly want more for less, is a challenge, but it can be done. Arguably, larger diversified companies – those with economies of scale and good cash flow – have the best opportunity for success. However, even smaller companies can be successful if management/ownership puts in place a good marketing strategy, watches cash flow (cash is still king, say financial gurus in the industry), and plays in markets/niches that tend to allow for greater profitability.

10.1 Learning from experience: what the observers saw in recent years

Plante & Moran PLLC (**www.plantemoran.com**) conducts annual surveys of the North American plastics industry.[1] Jeff Mengel, who heads up the Plastics Industry Team, analyzes the survey then presents the results at Plastics News' Executive forum each year. The survey typically analyses data from less than 200 companies, and not all are injection molders, but they represent a good cross-section of the industry.

Looking at feedback relating to 2008, Mengel discovered that **there was no correlation across the board between volume and profitability**. However, there was a negative correlation between profitability and volume when companies operated with fixed material cost contracts. Automotive customers are notorious for rejecting material price adjustments, and consequently the molders that supply automotive generally had 30% lower profit margins than other industries (the only industry that statistically correlated with lower profits).

Productivity improved as measured by value add (sales less material, purchased components and other outside processing). Value add per employee (all employees) rose to $76,486 in this survey, but the bulk of the productivity improvements were used to pay for an increasing cost of labor.

Additionally, Mengel noted that **machine utilization does *not* correlate with profits**. Utilization is a magnifier of the value proposition, Mengel explained. If you compete on price, the market dynamics of supply and demand have compressed your margins as there are just too many molders willing to quote using 'survival' pricing. *"Higher utilization is required just to retain your desired profitability,"* he said. *"However, if you compete on a unique solution, your margins are higher and higher utilization creates higher profitability."*

Sales growth reported in the survey was surprisingly low. Average two year sales growth for 2008 was only 1.7% with only 51 companies having more than 5% two year sales growth. Again, surprisingly, growth does *not* correlate with profits. *"Unfortunately, some companies are not strategic regarding the type of growth they desire or how they will accommodate the growth – and they suffer as a result,"* Mengel reported. *"In general, high-growth companies incur higher debt and interest charges and have higher overhead to digest the growth. Toward the end of 2008 and into 2009, any growth was rare and these warnings of non-strategic growth became hollow. Our current concern is the desire to increase sales at the expense of appropriate margins."*

---

1 North American Plastics Industry Study, www.plantemoran.com/industries/plastics/north-american-plastics-industry-study/pages/home.aspx.

Plante & Moran's next survey (released in March 2010, describing 2009) focused on four areas: sales growth and recovery, human capital, continued cost reductions and credit management. Despite the 'turbulence' of the recession, fewer than 10% of the respondents indicated that they were still in a phase of contraction. The majority (66%) viewed themselves as surviving/maintaining in 2009, had strong debt to equity ratio of 2:1 or less (56%), and a strong, or at least unchanged, relationship with their bank (72%). Most (72%) believed that the economy would take two or more years to get back to normal. However, in contrast to the previous year, companies had typically reduced labor costs by now, either by cutting rates or headcount.

Tight credit had reportedly caused the customers of these molding company respondents to reduce purchase orders and to start shopping around for better deals in order to preserve cash (66% of respondents reported moderate to high levels of new quoting activity). This resulting loss of revenue by plastics processors was driven more by volume reductions (71% lost more than 10% of sales due to volume reductions) than by customer bankruptcies or losses to competitors (29% lost more than 10% of revenue due to these factors). However Plante & Moran commented that it should be easier to replace lost revenue due to volume than to replace lost customers.

Mengel noted in his analysis that once again, **diversification played a big role in the success or failure of plastics processors during 2008 and 2009**. *"Those who failed to diversify their customer base suffered the most (22% of respondents lost 15% or more of their revenues due to customer bankruptcy or vendor switching),"* he stated. That resulted in plastics processors doing their own cost cutting and the first area was reduction in the labor force, with 33% of respondents reporting that they reduced the headcounts by more than 10%, and 31% having reduced their direct labor workforce by at least 25%. Interestingly, the survey showed that 27% indicated they did not reduce their workforce at all.

2010 survey responses (analyzed and published in 2011) showed some good signs of recovery. Press utilization increased, although it is still low, and this is a good sign for those companies achieving good margins. Average labor costs dropped slightly, and so did the cost of resin as a proportion of part price, as molders were able to pass resin price increases on to customers. Personnel turnover reached an all-time historical low for this survey, with the best, most experienced staff being retained.

On the downside, growth opportunities were limited, with customers industries such as construction, furniture and industrial products all still struggling. The auto industry experienced a recovery, but certainly not to pre-recession levels. Another area for concern which was flagged up this year was the continuing low level of investment in new production equipment, with the result that the average age of equipment continues to increase. Competitors in China for example are investing a great deal more in new technology.

10.2 Lessons learned and progress toward profitability

OEMs are typically uncomfortable with insecurity among their injection molded parts suppliers. Many were caught having to move tooling overnight from a plant in which the doors were to be padlocked, which makes the OEMs wary of these typically small, privately held businesses. Currently some OEMs are even beginning to look at the profit and loss (P&L) statements of suppliers like small molders, and performing other due diligence before deciding to place work with them. Caution reigns.

Conversely, many molders and moldmakers are taking a tougher stance with their OEM customers as well, particularly in the automotive market. While few resin suppliers are willing to even consider contract pricing of large quantities of material, molders are forewarning customers that their piece part pricing will rise in accordance with the cost of material. And they are telling them – not asking. It's no longer a choice for molders. They have to be paid what the parts cost plus profit or the OEM will find even fewer suppliers from which to choose. Moldmakers in particular are demanding – and getting – better payment terms that keep cash flowing.

So, to survive in the current tough climate, **every molder needs to be prepared with a strong business plan and a base of strong, diversified customers who understand and will pay for the value that the molder brings to the table**. As Plante & Moran's survey clearly showed, diversification is key to minimising the impact of a downturn that can affect one industry or one customer in that industry. Many custom molders understand this so well that they limit the percentage of business from any one customer to 20%. Similarly, a number of large OEMs, knowing the temptation that getting a lot of work from one big customer can be to a molder, limit their business to a particular molded parts supplier to no more than 30%. They don't want to be responsible for the demise of a supplier should their own business take a downturn.

> ☞ **A molder can be extremely busy – operating at near 100% machine utilization – and still not be profitable.**

Many molders can remember the 1990s when business was extremely good, and press manufacturers were backlogged six to eight weeks on certain size presses.

As a result, since business has slowed, there has been a tremendous amount of used equipment on the market in both North America and Europe. Many good used machines – some with very few hours of use could be had for a fraction of the original price.

With **capacity utilization** at 50-60% during much of the first decade of this century, many molders learned a hard lesson with regard to over-buying equipment. Some took the simplistic view that a high capacity utilization rate is an indicator of success – that is the more of their machines they have running, the more profitable they are. As the Plante & Moran survey noted, capacity utilization does not correlate with profitability. A molder can be extremely busy – operating at near 100% machine utilization – and still not be profitable.

Mengel believes, and it is borne out in many custom molding plants, that a healthy utilization level for the custom molding industry is a maximum of 75% due to the complexity of the press array required. This much flexibility is needed in order for molders to be able to respond to customers' demands and to be able to provide a wide range of press sizes to accommodate a wide variety of part/mold sizes. Thus, not all presses will be busy at the same time.

The temptation for molders in the headier days of the 1990s was to purchase a new press every time they got a new customer/job, believing that because all the presses were running they didn't have additional press capacity for more work. That decade expanded capacity to an exorbitant level, and with the subsequent drop-off in business it took nearly all of the first decade of the 21st century to haul capacity utilization back up around the 60% mark. In a speech in 2008 Jeff Mengel commented that another 85,000 presses could be unplugged in the US just to get capacity to the desired 75% level.

> ☞ **A healthy utilization level for the custom molding industry is a maximum of 75%.**

10.3 Operations software

Perhaps some of the problems encountered with press over-capacity come as a result of molders not truly knowing or understanding their production capacity. Scheduling can become a nightmare, and **it's still a common problem for molders to run short of material and have to stop the press to order more**. Material planning and scheduling have seldom been a molder's strong suit.

There are many helpful software programs that can perform integrated functions from the shop floor to the front door. Today, these are called ERP systems for

Enterprise Resource Planning. A couple of decades ago, it was known as MRP or **Material Requirements Planning**, because the only thing that manufacturers were concerned about was ordering in material, putting it in inventory, and knowing how much would be used on jobs. A min/max system would be used to trigger new material buys. For example, when the clear polycarbonate for a specific job got down to one storage bin, that would be the 'minimum' trigger to whoever was responsible for purchasing that more material had to be ordered.

Today, ERP systems can cut across the entire company and all the operations are integrated so that typically material purchases trigger accounts payable, and purchase orders are automatically matched up with payments. When parts are shipped, the information is recorded and automatically sent to accounts receivable, and the shipping packing list may trigger an invoice to be sent to the customer. Many ERP systems have quoting applications for quoting parts when they are ordered; engineering information; job information; labour, machine hours, quality information and any number of manufacturing plant applications.

IQMS (**www.iqms.com**) produces an ERP software package specifically for plastics molders. The company's roots are in plastic molding, so its system is specifically geared to the needs of molders. Useful features include real time production monitoring that automatically tracks data directly from work centers and updates production schedules and inventory instantaneously, multi-tool and family tool functionality that can handle multiple part numbers within the same tool, regrind usage and consumption features and more. The latest version is also written to enable users to comply with International Financial Reporting Standards (IFRS).

Software from Exact JobBoss (**www.jobboss.com**) is designed specifically for the job shop, which includes molders and moldmakers, and provides an integrated package to handle the one-off types of jobs in which mold manufacturers are typically engaged. Exact JobBoss contains a full product suite that includes quoting, scheduling, job tracking, costing, purchasing, inventory, shop floor, data collection, quality, customer relationship management (CRM) and integrated accounting. Made2Manage (**www.made2manage.com**) is another example of ERP for small to mid-sized manufacturers, and it also manages the entire business environment from selling and design to manufacturing, distribution, finance, and customer service and support.

One critical area that these ERP systems handle is job costing, which is critical to molders. Unfortunately, some molders still do not know their cost-to-manufacture, so having a computer data-based system can help them to track costs and prevent losses, or to at least pinpoint where the losses occurred so that it doesn't happen again.

Selecting and implementing ERP

Inside-ERP, a company that helps users evaluate and select ERP software, offers a free buyers' guide to mid-market ERP on their website (**www.inside-erp.com**). They have also put together a white paper listing important questions that you should ask yourself when selecting an ERP software program. Some of these questions are more applicable to larger organisations, but they all highlight important issues when implementing changes to the way you run your business. Investment in ERP software will only be cost-effective if you can identify the critical areas of your business, and you are willing and able to make the necessary changes.

- Do your organization's leaders support the ERP implementation project? This is key because most major operations systems need top down motivation to get the rank and file to accept the change that comes with systems.
- Who are the managers who can be responsible for measuring the business benefits relevant to their department's ERP modules? It will take more than just the IT person to make this successful. Every manager in every department such as quality, scheduling, material purchasing, and production, impacted by the ERP system must 'own' their part.
- Who will be the ERP project manager? Implementing the ERP system needs a 'champion' – someone who 'champions' the system and its implementation.
- What are the specific business problems you need to solve with ERP? Inventory shortages that shut down a press? Do you need shorter lead times on critical resins?
- What are the goals and metrics that you will use to measure the business benefits of your organization's new ERP solution i.e. the key performance indicators such as inventory accuracy, cost reductions, etc?
- What features and functions do you need from a new ERP solution that will help increase users' productivity and provide access to the business data users most need?
- Are there best practices you need to adopt with the ERP implementation?
- With which modules will you begin rolling out your ERP package? Many systems are modular such as JobBoss, Global Shop Solutions and IQMS among others. Evaluate which modules you need to start with – maybe the most critical area of your business like tracking inventory, purchasing and usage.

- Is the prospective ERP solution built with SOA (service oriented architecture) capabilities? This allows for greater flexibility and agility in the IT environment, connecting systems and automating manual business processes.
- Which users across your organization will need to be trained on the new system? Training is key! Don't assume that everyone will know what to do from the get-go.
- Will the ERP package be able to adapt to changes in your business as your company grows? Again, modularity is important so that add-ons can be implemented as the business changes.

10.4 When big sales look tempting

One of the lessons learned from the 1990s is that **sales cost money**. Many custom molders do not track their cost of sales, and thus don't have a good handle on what it actually costs to get and serve customers. While the giant molding job that comes with a dozen molds and millions of parts annually looks tempting to the average molder, who often dreams of such a big job, it can sink the ship.

Inventory alone can cost a tremendous amount, and if it's not turned over on a regular basis – several times a year at least – having that material laying around eats into the company's revenues. Molders, especially those that specialize in large parts, can be on the hook for purchasing tons of material that needs to be paid for in 30-60 days – if your material supplier is kind. In some cases, the molder can't afford to purchase the required amount of material, so the OEM will purchase the material and have it drop-shipped to the molder. That's just the material. If there are any value added operations such as hot stamping, sonic welding or assembly, that might also require special purchases of inserts, inks, foils etc., and the costs can go even higher.

In the Plante and Moran surveys discussed earlier it has been noted over the years that **companies with a highly complex mix of business (lots of molds, lots of materials, lots of customers, and lots of secondary or value-add operations) have a more difficult time keeping the bottom line in the black**, than companies with a lower mix of business. There is clearly a balance to be found between profitable diversification based on strengths, and attempting to be all things to all people.

Plante and Moran's Jeff Mengel has also noted that **successful companies' new sales possess a higher value proposition than the current sales**, so margins improve year on year, even with negotiated price reductions. Successful companies also maintain excellent pricing disciplines with very few sales with less than 10% margins. Most importantly, they retain slight superiority in just about every cost category – the cumulative impact having a positive impact on the bottom line.

10.5 Margins

> ☞ **The more commoditized the parts and the capabilities you offer, the less money you can get for what you do.**

Looking more closely at margins, these are typically thin, and **the average custom molder is lucky to get 10-12% margins on most jobs**. However Plante & Moran noted in their report that highly successful companies in 2008 had gross margins exceeding 28% – a far cry from the industry average. This cannot be achieved simply by minimising costs: *"Highly successful companies invest more in sales and market and have a broader diversification of customer. These successful companies exploit a niche and possess a value proposition in technical skills."*

Thus one of the benefits of having a specific niche in which there are fewer players, or serving specific markets in which not everyone wants to play, is that you can get higher margins for the work. Molding companies with unique capabilities (molding extremely tough thermoplastic materials for aerospace, for example) or perhaps offering compression molding along with injection molding for unique assemblies, can command more money for the parts they produce.

The more commoditized the parts and the capabilities you offer, the less money you can get for what you do. The same is true for markets you play in – the more molders that are also playing in those markets, the more competitive the price which translates into thinner margins. 'Me too' molders have a tougher time proving that they are different from all the others.

For example, StackTeck Ltd., a Canadian moldmaker (**www.stackteck.com**), installed an in-mold labeling (IML) molding system in order to offer customers the company's expertise in building molds for the IML process, and to perform qualifications on the molds in-house. StackTeck adopted IML technology as a niche expertise, and even received a patent on a new mold design that permits the ultimate in thin-walling for food packaging called the TRIM technology (Thin Recess Injection Molding), in which the section covered by the label was thinner and the label becomes part of the wall section of the container.

Some molders extend their geographic reach as well as their capabilities in order to expand business and improve margins. Mergon Corp., headquartered in Westmeath, Ireland, is a custom industrial blow molder that specializes in medical, industrial and automotive (**www.mergon.com**). In 1998 the company set up operations in Anderson, SC, to serve the North American marketplace more conveniently, and extended its reach still further by setting up in the Czech Republic in 2004.

In 2008, Mergon also expanded its technical capabilities into injection molding, increasing opportunities to put value in and take cost out. This was a good fit for Mergon because the company blow-molds a lot of components that are married to injection molded components. By extending its reach into a complementary process, Mergon was able to provide more value and thus increase its margins. It was also able to reduce costs because the company no longer had to ship blow-molded parts via a custom molder for completion, which saved on warehousing and shipping, and improved delivery times.

10.6 Firing unprofitable customers

Jeff Mengel of Plante and Moran has noted more than once in his evaluations of the plastics industry that not all customers are created equal, and not all customers are desirable. *"It's important to know which customer relationships you want to protect and which customers you can be more aggressive with,"* he said. *"Customers whose parts involve a high degree of complexity but low contribution to the company's profitability might be ones to weed out."* He suggests that molders go through their customer list in the same manner they do their financial portfolio, evaluating the contribution of each one to the overall value of the portfolio and making changes where needed.

Having completed your strategic marketing planning as described in earlier chapters, you should have a profile of the ideal customer for your company. Based on this information you can review your current customer list and identify those who are no longer a fit. Just as your business changes, so the ideal type of customer for your company will change as well. **Often, a customer that was a perfect fit for your business when you started out twenty years ago will not be a good fit today**.

Sometimes when business get slow it may be necessary to take on almost any type of work just to keep the presses running. However, when times improve and you begin getting business from more desirable, target customers, you need to ensure you have the capacity to serve them.

☞ **Not all customers are created equal.**

That might mean firing those customers that use more resources than they are willing to pay for or that are no longer profitable for you. A customer ordering 10,000 parts a year with 2,500 shipped each quarter might have looked pretty good when you were starting out or when business was slow and you had a lot of open press time. However, if you now deliver mainly long-running, high-volume jobs, getting that low-volume mold into a press could mean interrupting a long and more profitable run.

Customers who are slow paying, or who demand a lot of quality criteria but who fight you on price increases, can be more trouble than they are worth. Firing them might be the best thing for your business. Actually, firing them is a bit harsh. You might prefer to gently help them find another molder more suited to the type of molding work they have, and offer to get the mold ready and take it to the new molder.

Sometimes customers threaten to pull their molds from your facility, especially when you have to raise the prices due to material cost increases, additional quality criteria, or high maintenance on a mold. Maybe they've been doing business with you for so long that they resist any type of changes in the business relationship, and feel that threatening you is a good method to get their way. Let's face it; you can't continue to do business with a customer when it's just no longer profitable. **It's better to let unprofitable customers put your competitors out of business rather than you.**

Taking a hard line with customers that are slow paying or resist price increases, or are otherwise awkward can be risky. There may be times when they'll call your bluff and pull the tools, so you have to weigh the risks of taking this kind of action. Are you really willing to lose this customer? What are the pros and cons of pushing a customer?

We've experienced some major downturns during the first decade of the new century, and while you might not think now is a good time to weed out the less than desirable customers, a downturn can be the perfect time in which to do this. They're using resources that you can use for more profitable customers when things pick up again. If your presses are full of work from customers who are slow paying or using more resources than they're willing to pay for, you won't have the press capacity to take on the really good jobs that will come along with things get better.

Patrick McNally from consultants Blackman Kallick LLP (**www.blackmankallick.com**) once suggested that one way a company can take advantage of troubled times is to fire those customers who cost a lot to keep.[2] *"Especially if your cash*

2 Making Hay in the Rain, September 30, **2003**, www.plasticstoday.com.

is tight, you need to take a hard look at your customers and make sure you're making money," McNally said. "*Consider raising prices and reducing terms on your marginal customers. If your customers are in trouble and can't go anywhere else, they're going to have to meet your prices and terms. And if they can't, let your competition have them.*"

10.7 Typical signs of success

Plante & Moran's Jeff Mengel has identified several specific traits amongst the highly successful respondents in their studies, which summarize those things that all molders should strive to achieve to assure survival and success:[3]

Traits of highly successful companies

1. The highly successful companies are not large. The technical niche they exploit may be better suited to a narrower market. But rest assured, they are not thriving on volume.
2. Highly successful companies have more working capital availability – more cushion to address capital investments and daily operations.
3. Highly successful companies are two times more likely to be component specialists than the average molder.
4. Highly successful companies may invest heavily in new equipment, but net equipment is still a smaller percentage of total assets than the average molder. More of the successful companies' assets are in receivables and inventory.
5. The highly successful companies have average utilization, but also above average manufacturing complexity, primarily due to the number of active molds required to match their higher degree of customer diversification. It is necessary to have available capacity to manage complexity and the vagaries of the customers' demand.
6. Highly successful companies user predominately intermediate and engineered resins, but also purchase a higher percentage of resins under the customers' POs (purchase orders), meaning they are not rewarded/ punished for commodity risk.

3 North American Plastics Industry Study, **2009**, www.plantemoran.com/industries/plastics/north-american-plastics-industry-study/pages/home.aspx.

10.8 Profitability checklist

Getting a grip on the following aspects of your business, and tracking them over time, will help to maximise profitability:

- ✓ **Materials costs**. Fixed cost contracts impact profitability. Are you passing price hikes on to your customers?
- ✓ **Added value**. The more value you can add per employee (sales less all bought-in materials and services) the more profitable you should be.
- ✓ **Margins**. Look at the margins on every job. Focus on those customers and specializations where margins are highest. Avoid 'survival' pricing.
- ✓ **Machine utilization**. This is only reflected in higher profits if all jobs are delivering a good margin. Some spare capacity is essential to offer a timely and flexible response to good customers. Permanently idle machines could be sold off to free up space and cash to invest in higher value operations.
- ✓ **Customer base**. Diversity protects you from sudden downturns when a business changes. A broad customer base in a range of markets can be good; but being 'all things to all people' is not.
- ✓ **Inventory and resources**. Too much inventory affects cashflow. So does running out of materials and failing to deliver on time! Look at resource planning, and the possibility of customers purchasing raw materials for bigger jobs.
- ✓ **Cost of sales**. Sales to new customers can be expensive to achieve, so make sure the margins are there.

Technological challenges and opportunities

Chapter 11

Advanced injection molding technology – adding value and flexibility

Technology in injection molding and mold manufacturing has come a long way over the past two decades, much of it driven by developments from OEMs that demand new ways of providing complex components, new designs to enhance competitiveness, brand awareness, and consumer preferences. Resin producers have contributed to this as well by developing new materials for new applications in a variety of markets. Many new molding techniques or technologies have come about in an effort to save money in manufacturing processes by increasing productivity or eliminating secondary operations.

At the time of writing, in spite of the general gloom in the manufacturing sector, some suppliers to the molding industry were reporting positive indicators. In particular, several hot runner suppliers reported improved demand in specialised areas such as multi-material applications (two-shot molding) and techniques for making complex parts, suggesting that companies able to add value and offer something different may be better placed to invest. **OEMs are pushing suppliers to reduce costs and cycle times and improve quality, and the successful molders are using specialised technologies to meet these demands.**

There are some molders that make one or more of these new or improved technologies their niche, and truly specialize in these. Other custom molders see the benefit in being able to offer a variety of molding technologies to give them more opportunities with their OEM customers. The technologies mentioned in this chapter, that have been developed to a high level over the past decade, can lift a molder out of the 'shoot-and-ship' category and offer much greater potential for profitability. Typically they can reduce costs through improved cycle times and elimination of secondary operations, or improve quality and product choice, providing OEM customers with more options and improved service.

11.1 Two-shot molding

Many processors are now molding particular parts containing two different materials in one mold, as opposed to using post-assembly or two molding presses. Prior to the two-shot molding process, one common way of combining two materials was to mold the harder material first, then move those parts to a second molding machine, placing them inside the mold cavity for overmolding with the second material. Two-shot molding allows the first material to cool before the second one is injected. Critical variables such as draft and mold temperature must be considered when using two-shot molding. Doing both operations inside a single mold creates an integral part instead of two separate molding operations. This not only makes for a better part, but reduces costs through the elimination of a second molding machine, and improves quality as the parts don't have to be handled as much.

Also called multi-material molding or multi-component molding two-shot molding uses twin nozzles and material hoppers to shoot two different materials into the mold at the same time and produces an assembly comprised of two or more integrated components. Typical applications for this technology include vehicle tail lights combining two different coloured materials. **The advantage of combining two parts into one is a reduction in assembly time and the cost of manufacturing**. That said however, mold costs are higher. Another common example is a toothbrush handle, where a TPE is shot over a thermoplastic substrate to provide a soft grip feature on the handle. There are many components that use this technology now, and the demand for two-shot is increasing across a variety of applications.

Figure 11.1

Two shot part showing two different materials/colors for an automotive part from ABA-PGT Inc. (www.abapgt.com)

In any process combining two materials that have different properties, there can be problems in getting the materials to bond. However materials technology is advancing and the range of possible combinations is increasing.

Information on two-shot molding can be obtained from a number of material and machinery producers. For example, Ticona Engineering Polymers offers helpful information including product data for its materials that are suitable for two-shot or 'multi-component' molding, particularly in hard/soft combinations. The materials that Ticona offers for this process are Celcom or Hostaform polyoxymethylene, Celanex polybutylene terephthalate and Fortron polyphenylene sulphide, with a variety of elastomeric materials recommended as the second component. Ticona's information on multi-component molding (**www.ticona.com**) provides advice on the process and other helpful tips in a technical paper titled 'Hard-Soft Combinations with Hostaform, Celanex and Fortron.' Another technical paper is available on their website titled 'Hard Rules for Soft-Touch Overmolding'.

KraussMaffei (**www.kraussmaffei.com**), a maker of injection molding machinery, specializes in two-shot molding presses and technology. The company developed a process, ColorForm, that it showcased at K 2010 in Dusseldorf, Germany, which they claim will "*revolutionize the production of parts with high-gloss surfaces*" and replace post painting operations. ColorForm combines injection molding and a reaction injection molding process to produce multi-component parts with high-gloss surfaces in a fully automated single-stage process.

The process can help molders realize energy efficiency and cost benefits, according to KraussMaffei's Frank Peters. "*Integrating the painting process into the injection mold results in a far more reliable process and saves costs in several ways. Because the parts are painted in the mold immediately after they are produced, there is no need for the cleaning processes required prior to conventional painting. In fact, the whole chain of logistics processes necessary for conventional painting falls away completely.*"

"*Unit manufacturing costs are reduced without sacrificing any flexibility. The thickness of the coating and the surface hardness can be set to suit part geometry and design specifications. ColorForm can be used to apply any color, from glass clear to piano black, including color management.*" The ColorForm process can be used with many different substrate materials including thermoplastics, thermosets and PUR.

For molders desiring to get into multi-shot/multi-component molding in a cost effective manner, there is an alternative to buying a two-shot press. MGS Manufacturing Group Inc. headquartered in Germantown, Wis., developed a proprietary product, the UMS (Universal Multishot System) portable injection unit. The UMS units hook directly to existing molding presses to make them multi-shot capable. They can easily be transported between multiple injection molding machines facilitating quick program changes. An independent hydraulic unit that sits next to the press provides power and control for all functions (www.mgstech.com).

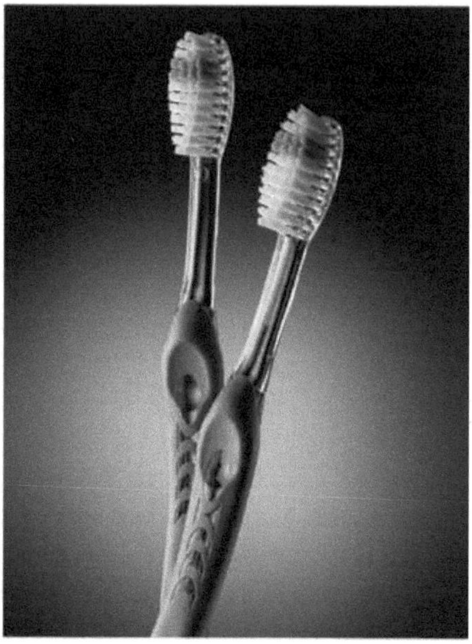

Figure 11.2

Two-shot toothbrush handles detailing the thermoplastic substrate with a TPE overmolded layer for ergonomic comfort and easy handling

11.2 Co-injection molding

This is similar to two-shot molding and is also called 'sandwich' molding, and again requires two nozzles that shoot two types of material into the same mold. The process involves the injection of a 'skin' to partially fill a cavity, followed by the core material to pack out the part. In this instance one material is effectively sandwiched between outer layers of another. Some molders use this method to **incorporate a recycled layer between two layers of virgin plastic, which helps to reduce the overall cost of the part.** This process can use two injection units and rotary molds designed for sequential injection, or a robot transferred mold, according to information from Ticona's website.

11.3 In-mold decorating (IMD) and in-mold labeling (IML)

This technology caught on quickly in Europe and has seen the most growth on that continent over the past decade. However, IMD and IML are also beginning to catch on in the US There are a number of factors that contribute to this growth disparity between North America and Europe, see section 7.4.

Figure 11.3

These food containers use StackTeck Matrix's proprietary TRIM technology (Thin Recess Injection Molding) for IML to reduce material usage while maintaining the container's integrity by using the molded-in label to provide extra strength (*www.stackteck.com*)

Innovative Injection Technologies, known as i2Tech (**www.i2-tech.com**), a custom injection molder and contract manufacturer, produces many parts using the IML technology. Molds built for these processes have to be carefully designed specifically for the process. i2Tech uses mold-fill analysis to ensure that the design meets the requirements. The mold not only has to be designed and built to accommodate the label but must accommodate the automation equipment such as the robotic arm to place the label in the mold. The pre-cut labels are stacked in a magazine, and then the i2Tech system uses static electricity to hold the label in place while the mold closes. Another key to designing a mold for IML is keeping the gate away from the label. According to i2tech, "*Labels can present a problem with gating because you need a uniform fill front. Also, you need good venting in the right area. Those are the two critical areas when building a mold for IML.*"

Typically, the end-of-arm tooling (EOAT) is purchased with the mold and designed to work together. In fact, designing an in-mold labelling system is typically a collaborative effort which can involve the label printers, the label makers, the robotics company, the injection molding machine company and the moldmaker. Today, labels have higher quality graphics and can be printed on almost any substrate material. There's also less handling with the in-mold process, compared with hand application of pressure sensitive labels after the part is molded, and secondary operations are reduced.

There is increasing demand for in-mold labels for durable goods such as lawn and garden products, and also recreational vehicles such as snowmobiles and jet-skis, where it is extremely important that the labels remain adhered to the product for safety reasons and due to product liability issues. Arctic Cat, one of i2Tech's customers, needed a permanent safety label on its snowmobiles and opted for in-mold labeling. The advantage of in-mold labels is that they are permanent – they actually become a part of the object. They are also typically made of the same material which can also be beneficial for recycling.

IML clearly requires significant investment. For fast-moving consumer products, Henkel's Shelly Gibson, purchasing manager for printed labels for North America, has commented that *"If you're only to be going into IML for a year, it doesn't work. Everything is going to the last point of differentiation, and IML makes sense with personal care products. But, it becomes a headache if you're not in it long term."* Keith Cardinal, senior packaging engineering manager Laundry Care for Henkel North America in Scottsdale, AZ, has also commented that Henkel uses a variety of labels, including in-mold, but adds that it's all about flexibility. *"One thing we need in the injection molding side is next generation robotics for label placement"* (**www.henkelna.com**).

David Coughlin is Director of Operations for Industramark, a division of Standard Register, (**www.standardregister.com/industrial**) which specializes in microporous film labels for both the packaging and durable goods markets. He has also commented that significant capital investment is required in any custom injection molding firm wanting to provide IML or IMD capabilities. The mold design and build has to accommodate the process, and the machine and downstream equipment all need to be integrated. Typically this will include robotics and either magazines for pre-cut labels or alternatively cut-in-place technology. There may also be a number of different suppliers to be coordinated.

11.4 Film insert molding

Film Insert Molding (FIM) is much like other types of insert molding and very much like IML/IMD. Like these processes it can be a versatile and cost effective method of decorating and manufacturing plastic parts. The flat film is first reverse-decorated (typically by a screen printing process), then formed (generally by thermoforming the rigid film layer) die cut, and placed into the mold cavity with the use of robotics before being back-injected. Using the FIM process, it is easy to integrate components such as a lens and body into a single unit comprising just one piece of a formable hard-coated film. For example, **mobile phones (cell phones) have become a major user of FIM for the manufacture of the front part of the case because the colored body and the clear lens over the screen can be molded in one part.**

According to MacDermid, a manufacturer of hard-coated films in the UK, FIM saves on the additional tooling that would normally be required, and the assembly costs. Additionally, products can be formed into almost any shape and depth. FIM technology is a good fit for a variety of end-use markets including electronics, telecommunications, appliances and automotive.

11.5 Stack molding and tandem molding

> ☞ **Stack molds can virtually double the output of an injection molding machine.**

Stack molding technology has been around for quite a while, and has become widely used to increase productivity. Many mold shops build stack molds. Stack molds have two or more parting surfaces, usually with the same arrangement of runners and cavities in each one. This enables processors to get double (or even triple) the volume out of a mold on the same press as a single face mold, avoiding the need to go to a larger press to accommodate a larger projected surface area for more cavities.

The larger the projected injection molding surface is, the larger the tonnage of machine is needed, whereas stack molds typically do not even require a significantly greater clamp force, although tie bar spacing is obviously a consideration. By molding in the same projected area and getting double or triple the cavitation, molders increase their volume and productivity.

Mold Hotrunner Solutions (MHS), a global supplier of hot runner systems for a variety of applications, builds special hot runner systems and valve gate controllers for stack molds. On its website, MHS notes that while stack molds are often overlooked as a viable option in the design phase because of a lack of familiarity among designers, stack molds can virtually double the output of an injection molding machine by distributing the plastic melt into two separate mold parting surfaces. Instead of increasing the mold size by adding more cavities onto a single-face mold, a stack mold simply adds another layer of cavities parallel to the first layer. According to MHS, the fill, pack and cooling time remain the same for a stack mold and only the mold open and mold close time will add to the cycle time (**www.moldhotrunnersolutions.com**).

Stack molds are generally used for flatter parts because the mold opening for each face is quite narrow. With a stack mold, a molder can maintain the same mold footprint and use the same press size, while producing twice or even three times the number of parts. For example, a two-cavity coffee can lid mold can, by using a stack mold, produce four or six lids using a double or triple-face configuration.

While more expensive initially, a stack mold improves productivity and efficiency to the point that the return on investment (ROI) is generally quick and quite good. However, remember that this is a technology that fits specific types of parts. If you are currently molding these types of parts in single-face molds, then stack molding will probably offer a good alternative.

Tandem molding is similar in principle to stack molding, but offers the potential for molding two significantly different components of a product on the same machine (such as a bucket and its lid). Developed by T/Mould in Germany (**www.t-mould.com**), the system has a unique locking system, which opens each of two mold faces in turn whilst holding the other one locked. Compared with running two presses, overall cycle times and hence cost can be reduced for suitable products.

11.6 Rotational stack molding

This intriguing technology enables the molder to do a number of in-mold operations. The Spin Stack technology was developed by Jes Gram, president of Gram Technology, who now licenses the technology to moldmakers and molders (**www.gramtechnology.com**). Like a stack mold, the Spin Stack technology offers the ability to mold a larger number of parts in one press, and provides the additional benefit of performing a number of additional in-mold operations such as labeling, painting, assembly, overmolding and other tasks. The mold rotates on a central core with each side rotating 90 degrees to a position where an operation is performed, then rotating another 90 degrees to a position where a second operation is performed, then another 90 degrees for yet another operation and finally to the position where the part is ejected.

For example, a thermoplastic substrate is molded in a 8-cavity mold, then the mold rotates 90 degrees to a position where a thermoplastic elastomer is overmolded onto the substrate, then rotated another 90 degrees where an insert is automatically put into each part, then rotated 90 degrees again before the parts are ejected.

A number of large molding companies have purchased licenses such as Nypro Inc. (**www.nypro.com**) and The Tech Group (**www.techgroup.com**), and also some mold manufacturing firms such as Tech Mold Inc., (**www.techmold.com**) a privately held mold manufacturer that specializes in building high-cavitation molds (some with 200 cavities or more) for the medical, pharmaceutical and packaging industries.

Other mold manufacturers have challenged Gram Technologies' patent, and created their own competing rotating 'cube' mold technology. Foboha GmbH in Germany, which is part of Advaltech (**www.advaltech.com**), has developed its own rotating 'cube' technology, and in July 2005 completed one of the largest projects consisting of eight 'stack-turning' molds for a customer in the consumer goods market. The 32-cavity molds were the largest 'cube molds' ever built, according to Foboha, with dimensions of 1,200 mm horizontal and 960 mm vertical and an installation height of 1,600 mm. They are used to mold 3-component deodorant roller parts on a K-Tec 450-3K from Ferromatik Milacron. The molds were designed to reduce costs and assembly requirements. They were piloted and supplied within just 9 months, and delivered a cycle time even shorter than specified by the customer.

Electroform Company Inc. in Rockford, IL, has also developed a version of a rotating stack mold, and showcased this at recent NPE shows in the US by making a toy car comprising a molded thermoplastic substrate, a decorative label and wheels, assembled in one operation.

While there are not an abundance of applications in which the Spin Stack or rotating 'cube' molds are required, where there is an application this can be an invaluable solution for an OEM looking to save time and money.

11.7 Structural foam molding

Structural foam molding is also called low pressure injection molding because it is similar to the 'normal' injection molding process for thermoplastics. However, rather than the high pressure process of injection molding, the structural foam process uses low pressure and relies on a foaming action caused by the injection of an inert gas distributed in the plastic material to facilitate the flow. Foaming can also be achieved through gases released by the decomposition of a chemical blowing agent added to the resin. This allows the molten resin to flow much further than the typical injection molding process would allow, at a much lower pressure.

According to information from Creative Techniques Inc., a manufacturer of innovative reuseable transport packaging solutions such as large containers, trays and pallets (**www.creative techniques.com**), most molds for structural foam are built from aluminum, which can be machined up to three times faster than tool steel. Consequently structural foam molds cost less at the outset. Also, by having longer resin flow lengths, parts can be molded that are much larger in size and weight than would typically be produced in a conventional injection molding process.

Structural foam molding is best suited to thick-walled parts, due to the lack of high pressure to fill the cavity. Often very large parts such as components for heavy plant, agricultural equipment and large truck bodies will use the structural foam process, where it has the added benefit of reducing weight and material costs. **Structural foam molding offers significant material density reduction, sometimes as much as 40% compared with solid material, while providing excellent strength-to-weight ratio**. Typical materials used for this process include high-density polyethylene (HDPE), acrylonitrile-butadiene-styrene (ABS), polypropylene (PP) and polyethylene (PE).

11.8 MuCell technology

Developed by Trexel GmbH, (**www.trexel.com**) the MuCell Microcellular Foam Process has found its niche across many markets in many applications. While the MuCell process is much like the structural foam process discussed in the previous section, MuCell was developed for smaller, thin-walled parts, and can help molders address issues such as unpredictable differential shrinkage (which can lead to sink marks, warpage, etc.). In addition to dimensional stability and quality improvements it is also claimed that the MuCell process results in reduced in-mold stress and the elimination of stress cracking. **MuCell reduces clamp tonnage and material requirements, which equals lower costs, reduced part weight and faster cycles.**

The MuCell process incorporates supercritical nitrogen in the injection molding process, which nucleates on entering the mold to create millions of nearly invisible microcells in the end product. This allows the thin wall cavities to be filled with reduced pressure and with reduced clamp tonnage, which permits additional thin walling. The expansion in the mold eliminates the packing and holding phase, improving dimensional stability, and the microcells replace their equivalent volume of plastic resulting in a cumulative reduction of up to 10% in package weight without any perceptible difference in the quality.[1]

The MuCell Microcellular process was used to produce a 500g margarine tub for Unilever, which won the German Packaging Award and a WorldStar Packaging Award in 2008. The tub is molded using SLIM (super light injection molding) technology developed by Veriplast Solutions (**www.veriplast.com**). This combines the MuCell foaming process and Veriplast's extra slim down-gauged in-mold label.

1 MuCell Process News, Vol. 18, February, **2009.**

11.9 Gas and fluid assisted injection molding

Unlike the foaming processes described above, moldings produced using the process described as gas assisted injection molding incorporate continuous hollow channels. Gas is injected into the mold cavity after the plastic to complete filling, generally reducing part weight, cycle time and pressure. Variations on the process can speed part cooling, and there is also a process called external gas molding, where gas is injected between a component and the mold wall, to improve contact with the opposite wall and the subsequent surface quality.

The process was first developed in the early 1980s but was the subject of high-profile patent disputes for many years. A more recent advance is the use of water in place of the gas (water assisted injection molding), which improves cooling still further in thick walled parts.

Parts generally have good surface finish, and benefit from the ability to include thick and thin sections and sculptured shapes, such as many automotive handles and related components. Other typical applications include TV cabinets and printer housings, where surface finish and weight reduction are both valuable, components including handles and larger panels for white goods, molded chairs, and many other household items. The good surface finish and the resulting ability to be painted means that parts produced by this route can replace some die-cast metal products.

11.10 Micromolding

Small parts are becoming big business, as micromolding is catching on in various industries, particularly in medical products and electronics. Small devices such as mobile phones and PDAs have become ubiquitous, and demand ever-smaller, lightweight components. Some custom molding companies that previously specialized in molding very small or miniature components have easily made the transition to micromolding. Some micromolded parts are so small the details in the parts can only be seen with a high-powered microscope. Others are parts that can be seen with the naked eye, but have details and features that are too small to be seen with the naked eye. Definitions of micromolding vary, but typically parts may weigh less than 1gram, or contain less than a pellet of plastic. **The value of the parts produced clearly has little relationship to the quantity of plastic they contain, but everything to do with quality**.

Typically micromoldings will be made from higher value engineering and speciality plastics, and precision and consistency will be extremely important. Micromoldings are generally components of more complex assemblies, and the ability to incorporate inserts on a micro-scale can also be a useful technology to offer. Some companies are also able to undertake 2-shot micromolding.

Plastic Design Corporation (PDC) is an established custom molder specializing in micromolding of medical components, as well as a micro moldmaker (**www.plasticdesigncorporation.com**). President of PDC, Mark Kinder has commented on the difficulty of measuring micro components accurately; hence the company uses scanning electron microscopes. Another of the downsides to molding micro parts is the loss of mechanical properties, according to Kinder. *"Below 20 thousandths (500 μm), you lose tensile strength. Heat transfer is a problem too."* Usually that is because the dwell time in the injection unit is very long considering the extremely small size of the part.

In 2010 Arburg introduced a new micromolding machine that operates using two screws which separate the preparation, dosing and injection of the material. A servo-electrically driven mixing screw in the pre-plasticizing section, installed at a 45-degree angle to the horizontal injection unit, ensures that standard resin pellets are prepared under ideal conditions. The molten material is then moved from the pre-plasticizing stage to the injection unit, via the injection screw, which is used purely for moving the material (**www.arburg.com**).

Some resin producers make materials specifically for micromolding, such as Topas Advanced Polymers (**www.topas.com**) with its cyclic olefin copolymer (COC). Many companies that micromold medical parts need biocompatibility because the parts are placed into the human body. Topas COC resins are ultrapure and biocompatible to accommodate that requirement.

To be in the micromolding business takes special equipment for producing the molds, molding inspecting and handling the parts, and talent and expertise to work with the machinery, particularly in the manufacturing of the molds. It's not a fit for everyone, but for those who are into small parts, micromolding is the next logical step.

The prefix nano is also frequently used in the context of plastics molding, but there are no nano-scale moldings. Nano-materials are extremely small particles used as fillers and reinforcements within plastics (and other materials), which impart particular qualities, such as mechanical strength, flame retardancy and barrier properties.

☞ **Small parts are becoming big business.**

11.11 Choosing your process niche

Because most custom injection molders use the standard injection molding process for thermoplastic resins, choosing a process niche can be beneficial to your business. But, rather than jump into something entirely new just because it's the latest 'hot' thing in the industry, you should look carefully at your current capabilities, both in equipment and employee expertise, and then see what opportunities there are to build on those. If you currently specialize in very small parts, and perhaps run machines of 40 tons or less, then micromolding might be the next niche that makes sense. A molder who injection molds very large parts might consider the structural foam molding process.

Specializing in a process can be just as beneficial to your business and marketing strategy as specializing in a specific market. With a process, you can potentially enter several markets that require the same process for their parts (maybe medical and electronics for example), thus diversifying your customer base and taking advantage of a process technology that can build your business.

Chapter 12

Factory organization and efficiency

12.1 More than molding machines

Injection molding machine technology has evolved in many ways over the past few decades and new methods of molding have developed. There are machines that have two-shot capabilities with double barrels, machines that can accommodate a wide range of molds including three-plate stack molds, Spin Stack molds, and all types of in-mold labeling and decorating, gas/water assist, MuCell molding, and more. Molding machines have got larger to accommodate the molding of large parts such as bumpers for automobiles. At the other end of the spectrum molding machines have also got smaller to be able to mold 'micro' parts that are so small you need a microscope to inspect them. Vertical molding machines operate in high-speed insert molding environments and reel-to-reel (used in many electronics applications) with automation to manufacture components and assemblies. There are a number of good opportunities in vertical molding as there are fewer molders in that arena.

Electric molding machines became very popular during the late 1980s and 1990s, particularly in clean room molding environments, as a way to achieve better molding consistency and repeatability. In fact, in June of 2010, Swiss molding machinery maker Netstal Group announced that they "*see no future in full hydraulic machines, at least for the top end portion of the market that the company serves*," according to a news report.[1] That determination means that Netstal will gradually end production of its traditional SynErgy line of hydraulic injection molding machines. Changing market environments also influence the types of molding machines that companies produce. In Netstal's case, it was a huge player in the CD/DVD molding equipment business, with 40% of turnover coming from that market in the year 2000. With the entrance of the MP3 and downloadable music direct from the Internet, Netstal exited that market completely.

There are many ways to acquire molding equipment, as the molding machine manufacturing arena has changed dramatically over the past decade.

---

1 Netstal plans to end manufacture of fully hydraulic range, June 11, **2010**, www.plasticstoday.com.

Consolidation through mergers and acquisitions has resulted in the downsizing of the machine manufacturing industry to fewer majors than in the 1980s and 1990s.

Overcapacity in the molding industry also resulted in a glut of used molding machines on the market, many of them relatively new – purchased during the mid-to-late 1990s and early 2000s when business was still quite good and company owners were confident of the boom times remaining. As molding companies downsized their operations with the fall-off in business, or even went out of business altogether, machinery auctions became a weekly event. Interested parties could bid online for this equipment and pick up some very good equipment very cheaply. Even eBay had some good offerings, and molders scouring the internet for good deals often found what they were looking for at bargain prices.

However, nothing remains the same for long. Throughout 2009, used equipment was in big demand in the US, eventually reaching a point where it was actually in short supply. *"Everyone wants used equipment, but much of the used molding equipment that was on the market has been picked over,"* said David Hausmann, owner of Elite Machinery Systems (**www.elitemachinerysystems.com**). *"We had 117 horizontal injection molding machines at one time, but now we have just 44 left. And only 15 verticals – about half of what we usually have,"* said Hausmann in a June 2010 interview. Interestingly he also observed that whilst capacity appeared to be returning amongst molders, they still wanted to buy used machinery. *"Fifteen years people used to tell me they didn't want used machines, saying 'we only buy new,' That's how a lot of these guys got in trouble – buying new machines all the time. Now many of these same companies won't even look at new – they're begging and holding out for the used equipment. Things are turning."*

12.2 Robotics and molding manufacturing cells

Most custom injection molding companies today have implemented the use of robotics in some portion of their manufacturing business. In the 1980s, robots were considered for use only in very large companies in very high-volume manufacturing operations such as automotive. Whilst it's true that the automotive industry, the aircraft industry and other large industrial operations led the way into the use of robotics, nonetheless by the 1990s custom molders in Europe and North America were catching on to what the Japanese already knew: **robotics are for practically every manufacturer**.

Many molders chose simple 'pick-and-place' robots or sprue-pickers, which were robotic 'arms' that could reach into an opening mold, pull out the parts or the sprue, and set them on a table or conveyor belt for transport to the next operation.

> ☞ **Reducing cycle times and eliminating 'secondary' steps could mean the difference between profitability and loss.**

These pick-and-place robots became very useful in multi-cavity molds because one robot arm could have four or eight 'fingers' (suction cups or other devices to grab the parts) that could pick off multiple parts and place them on the table or the conveyor belt in one smooth motion. However, these robots were often programmed for just one mold on one press, so that when that particular mold wasn't running, the robot sat silently, its arm raised into the air as if waiting for permission to ask, "when am I going to work again?"

As robots advanced in capabilities along with ability to program them easily for different jobs, they became more useful in molding operations. For instance in an overmolding operation, they could grab one part (the substrate) from the bottom half of a mold while at the same time grabbing the overmolded part from the top half of the mold, then place the substrate part in the top half of the mold while placing the overmolded part on a table or conveyor belt. With global competition becoming vastly more 'competitive,' reducing cycle times and eliminating 'secondary' steps, particularly those performed by human operators, could mean the difference between profitability and loss.

Robots are also being used more and more on smaller presses, and for work that humans can do less efficiently such as placing inserts into cavities. For example, a robot can place all eight inserts into an eight-cavity mold in one motion. It would take a human more than eight times as long to place the inserts by hand. Additionally, robotics can reduce scrap rates by as much as 5%, say some robotics makers.

More and more molders are finding it efficient and more cost effective to organise their molding operations into manufacturing cells, which integrate a variety of robotic activities including secondary operations, or allow these activities to be performed near or at the press, reducing cycle time and labor costs. Quality and consistency are also likely to be improved. Typically a molding cell might also incorporate a conveyor and an assembly operation or another type of secondary finishing operation.

Robots are particularly efficient in large-part handling, and can reduce or eliminate injury risks to workers who might otherwise be required to handle large parts, such as those for automotive, industrial, building and construction or heavy truck markets.

While robots have helped reduce the need for full-time human activity at the press, requirements for people to perform robot integration, re-programming and maintenance have increased. With the skilled-labor shortage that most companies are seeing in North America and Europe, robotics have at least given molding companies some breathing room when it comes to the need to hire more people.

12.2.1 Summary of potential benefits of incorporating robotics

- Reduced cycle times
- Reduced labour costs, and addresses shortage of skilled labour
- Reduced injury risks
- Improved quality and consistency, reduced scrap rates

12.3 Secondary operations

Secondary operations such as hot stamping, sonic welding, assembly and other after-molding tasks that need to be performed to complete the product, are often done in rooms that are set apart from the main molding production floor. More molders are investing in secondary operations equipment so that they can perform more tasks downstream and therefore provide more value - and add more profitability to the bottom line.

12.3.1 Welding

Some of the most common secondary or post-molding operations performed are in the welding arena. There are a number of welding types including vibration or friction welding, contact welding, contact welding, hot-plate welding, high-frequency welding, ultrasonic welding, spin welding, laser welding and solvent welding. The purpose of any of these types of welding is to join two pieces of plastic without using adhesive or mechanical fasteners. The joining surfaces are softened and fused, and the key to this is that in most cases, the plastic materials must be of the same type.[2]

There are several companies that make the equipment for doing the various welding jobs, e.g. Branson Ultrasonics (**www.bransonultrasonics.com**) and Dukane (**www.dukane.com**). Technologies have evolved over the past three

---

2 Goodship, V.,The Instant Expert: Plastics, Processing and Properties, **2010,** Plastics Information Direct, pp 104-105, 978-1-906479-05-3.

decades, as have materials. Choice of the best welding method depends on factors such as the type of material, the size of the part, the requirement of the part (i.e. does the part require a structural weld?), and whether a water- or air-tight seal is required.

In 2009, Dukane introduced a all-electric ultrasonic press, the iQ Servo Welder that offers 'better repeatability' than a pneumatic press, according to Dukane. The iQ Servo Welder optimizes the collapse speed during the weld and hold phase, creating stronger pull test results. Collapsing too slowly allows for material degradation, while collapsing too fast may cause cold forming, according to Dukane. Matching the collapse speed during the melt phase of the process is critical to producing superior bond strength.

One of the newer methods of bonding plastics comes from Emabond (**www.emabond.com**). Emabond is an electromagnetic bonding method, and the company provides complete turnkey systems that include assisting with the design of the product to be welded and then designing and building the equipment for the customer. The Emabond process effectively combines elements of welding and adhesive bonding, using high-frequency electromagnetic radiation together with a 'susceptor' material such as a polymer resin placed within the joint. The process can be used to join dissimilar plastics, and also rigid to flexible parts.

12.3.2 Decorating

Decorating plastics can be a profitable area of secondary operations, often called 'value-added' activities, that provides a molder's customers with a finished product. Many products require logos, instructions or decoration, for product differentiation or to more permanently place warnings on specific products for example. Rebling Plastics (**www.reblingplastics.com**) is an example of a custom molding and moldmaking company which performs many different types of molding processes and secondary operations. Rebling offer two main ways to decorate plastics: hot stamping and pad printing.

Hot stamping offers a more expensive way of applying decorations, logos or critical warning verbiage that must remain on the product permanently. The hot stamping process uses a heated die which applies pressure to a hot stamping foil that is pressed against the plastic part. These hot stamping foils come in a variety of colors depending on what the customer wants. The heat and pressure from the die then transfers the pigment from the hot stamping foil onto the plastic part. Hot stamping can be performed on flat surfaces as well as round and oval surfaces because the dies used can be manufactured to accommodate various shapes of plastic parts. In fact, round parts can be decorated around the entire circumference of the part, or parts can be decorated on the front and back in one operation using a roll hot stamping machine.

Pad printing is a similar process, but offers a lower cost method of decorating plastic parts. An engraved plate, commonly called a 'cliché', contains the etched image of the required artwork. Ink is spread onto this plate and then removed, leaving the wet ink only in the etched areas. A silicone pad picks up the ink from the etched areas and transmits the image to the plastic part. One advantage of pad printing is that the image can be transferred to irregular surfaces since the silicone pad will conform to these surfaces during the transfer of ink.

12.4 Clean room manufacturing cells

Clean room equipment used in molding operations generally tends to involve all-electric molding presses to prevent oil and grease from getting into the clean room and contaminating the cosmetically sensitive parts or medical and laboratory components. If hydraulics are used, the 'working' end of the injection molding machine (the motor, hydraulics, etc.) is installed so that it is outside the clean room wall.

Most 'clean room' custom molders typically have Class 100,000 clean rooms, known as ISO Equivalent Class 8, or sometimes Class 10,000 (ISO 7). The differences in the levels of clean room classifications relate to the cleanliness required, for example requirements for the Class 10,000 clean room include a gowning room, hair and beard nets to be worn, sticky mats outside the production room door or shoe covers, and hepa (high efficiency particulate arresting or high efficiency particulate air) filters to eliminate dust.

More molders are entering the medical device molding market, which means more have implemented clean room molding operations. The use of the term 'Class 100' or 'Class 1000' refer to FED-STD-209E and denote the number of particles of size 0.5 microns or larger permitted per cubic foot of air. Ordinary room air is approximately Class 1,000,000 or ISO 9. The lowest classification is Class 1 or ISO equivalent 3; Class 10 is ISO 4; Class 100 is ISO 5; Class 1,000 is ISO 6.

Customers will often request 'clean room' molding for their components, but some ultimately baulk at the extra costs it adds. What some molding companies are finding out is that a less stringent 'clean environment' often fills the bill just as well, and at a lower overall cost-to-manufacture. There are alternative ways to establish 'clean environment molding' areas either with new equipment that is available from makers of clean room molding cells, or molders can build their own.

Case history: Mastercraft Companies

One such example is provided by Mastercraft Companies (**www.mastercraft-companies.com**), a group incorporating a moldmaker and a contract molder (Polycraft Industries) serving a customer base that is 60% medical. To accommodate some of those customers, it does have a clean room in its molding facility, containing five injection presses from 35 to 110 tons. However, fully certified clean rooms in the range of Class 1,000 are expensive to install and maintain. To meet some customers' lesser requirements for clean environment molding, Mastercraft began enclosing some of its presses in an acrylic shell, complete with hepa filter positive pressure air filtration system.

Mastercraft's goal was to offer its customers a cost-effective, yet approved method that follows good manufacturing practices for medical device molding. The Clean Environment Manufacturing Cell (CEMC) was the answer. The first three enclosures that the company installed were kits that the company purchased and assembled. The enclosures encase the entire press, conveyor system, and vertical top-entry robotics. Whenever an operator was required at the press, they would enter the enclosure through a door.

A second phase consists of a modified enclosure system that Mastercraft designed and manufactured itself from Plexiglas acrylic sheet and aluminum support structures. The modified enclosure is smaller and encases the back half of the molding press, the conveyor system, and the robotics. The company designed these smaller Clean Environment Manufacturing Cells to make mold and machine maintenance easier. Large units must be pulled away from the machine to change the mold. Also, the company found it gets better air control in the smaller units. To reduce the overall height of these modified cells, Mastercraft opted for side entry robots for parts removal. The modified cells also have hepa filter positive pressure air filtration, and an entry door to accommodate maintenance. Mastercraft say customers have accepted this new Clean Environment Manufacturing Cell quite well: "*With the modified Cell concept, using the side entry robot we use one third the cubic footage we used in the first three cells, and we're cleaning two-thirds less air, making the modified Clean Environment Manufacturing Cells much more cost effective.*"

Another advantage to the modified cells is that no crane is needed to change the mold, whereas the Phase One enclosures required a boom on a fork lift to remove a mold and to set another mold from the top. Because the front of the molding press can now be opened, molds can be removed and re-set easily and quickly. The newest Clean Environment Manufacturing Cell at the time of writing is designed for a new 143-ton Arburg Allrounder 520A. Currently, modified cells have been installed on five machines - 100, 110, a 143 and two 140 ton presses, with more anticipated.

The benefit to the company in these CEMCs is that there is no massive infrastructure required as there is for actual clean rooms, and the cells are portable – they can be picked up and moved to another press if needed. Mastercraft say that the Clean Environment Manufacturing Cells are key to the company's ability to meet quality standards while minimizing capital equipment investments, and lowering overall costs-to-manufacture.

12.5 Preventive maintenance saves money, extends equipment life

☞ **Processing machinery that is properly maintained typically delivers decreased energy consumption, increased productivity, better part quality and reduced downtime giving the equipment greater availability.**

Lack of maintenance on processing equipment can cost a company a lot in energy, processing inefficiencies and downtime. Keeping up with your plant's preventive maintenance (PM) on machinery can pay big dividends in the long run as many molders find out. In fact, operational efficiencies and overall costs to manufacture are tied closely to processing equipment. How well a processor maintains the equipment often equates to how well the company does at the bottom line. In a recent Aberdeen Group survey[3] it was reported that 89% of Best-in-Class manufacturing companies responding to the survey optimized their manufacturing operations through Overall Equipment Effectiveness (OEE), while still satisfying the energy, environment and business needs with respect to attaining sustainability goals in the company. (OEE is a composite metric accounting for availability, performance and quality.)

While few people think of machine maintenance as connected to 'sustainable production,' the Aberdeen survey showed a direct correlation between OEE and sustainable manufacturing goals. Processing machinery that is properly maintained typically delivers decreased energy consumption, increased productivity, better part quality and reduced downtime giving the equipment greater availability. Aberdeen Group's Sustainable Production report revealed that its 'Best-in-Class' respondents are integrating sustainable features across their operations, "*taking into consideration information related to energy, emissions, product, process and employee safety – while optimizing production, maintenance,*

3 Shah, M., Sustainable Production: Good for the Plant, Good for the Planet, September 30, **2009**, www.aberdeen.com.

and engineering processes." One example the report cited is including energy as a major factor in addition to asset condition while scheduling maintenance.

Looking more specifically at plastics molding, Paul Caprio, president of major machinery manufacturer KraussMaffei (**www.kraussmaffei.com**) said at the beginning of 2010 that while KraussMaffei's customers are concerned with energy consumption, this is still secondary to precision and output. "*Even though energy saving is a buzz word, quality parts and high productivity are what counts on the manufacturing floor,*" says Caprio. "*Our customers are more attuned to making the machine run better.*" That said, Caprio adds that many customers *are* getting rid of older molding machines that tend to consume more energy, and buying new or newer machines – particularly in the hydraulic models – that have more energy efficient motors and pump systems.

Machine maintenance has always been recognized as critical to quality parts and high productivity, but one thing that has changed, notes KraussMaffei's Caprio, is the greater reliance upon the machine OEM to supply that maintenance. "*There's no question that over the past two years, the customer base is absolutely relying on the machine manufacturer to handle the preventive maintenance on the equipment,*" says Caprio. "*Many processors have gotten so lean in laying people off and other cost-cutting measures that they're not doing machine maintenance in-house. One of the safety valves Krauss-Maffei offers is planned regular checkups, quarterly, semi-annually or annually.*"

Jimmy Langsdon, president of Columbia Machine Works Inc. (**www.columbiamachineworks.com**) is in the business of machinery maintenance and repair, including injection molding machinery. He says that many times, especially when the economy is slow, companies get too lean. "*A lot of people make the mistake of cutting their maintenance budget when dollars get tight,*" says Langsdon. "*This is the worst thing you can do. Keeping your machinery in tip-top shape is the most important thing to your business.*"

So important in fact, that Scott Paulson, president of Paulson Training (**www.paulson-training.com**), says his company offers a variety of injection molding training programs that include machine maintenance. "*Maintenance is part of the efficiency equation. Maintenance affects downtime - which is a killer - so it's vitally important,*" Paulson says.

☞ **Many processors have gotten so lean in laying people off and other cost-cutting measures that they're not doing machine maintenance in-house.**

> ☞ **When times are slow that's the best time to perform preventive maintenance on machinery, and use the employees productively.**

Langsdon says the primary problem he sees with molding machine maintenance is the tendency to cut obvious corners, such as lubrication. *"We see a lot of molding machine failures due to poor lubrication, either a lack of lubrication or dirty lubricant,"* he comments. *"When we rebuild injection molding machines, a common thing we see is bearings that are worn out due to lack of lubrication."*

Andy Routsis (**www.andyroutsis.com**), has made similar points, and also offers training in machine maintenance. There's definitely a huge payoff in maintaining molding machinery, and often it's the areas that people don't think about that can cause the biggest problems, like lubrication. *"They don't lubricate the machine properly – that's always a problem. Another big one is leveling, particularly in the larger-size presses like a 400 ton running an 8-second cycle – that machine is rockin' and rollin'"* emphasizes Routsis, who recommends a bubble level or digital level, or better yet a laser level to check the level of the machine and re-level them periodically.

Routsis also recommends checking the platens for simple flatness, something that he finds people rarely do. *"Take a 12-inch or 36-inch steel ruler and lay it across the stationary platen by the locating ring,"* he suggests. *"If you see a gap between the platen and ruler you know it's not flat."* Additionally, he adds, check the moveable platen supports. *"The reinforcements on the platens should preferably be cast not machined to last longer and require less maintenance."*

Another significant but surprising problem that arises on new equipment has to do with the laws regarding shipping a machine. *"The machinery maker will do a run-off to test all the functions of the machine before shipping to make sure it's running correctly,"* explains Routsis. *"One of the things they must do by law is empty the machine of oil. From the run-off, you get a lot of chaffings from the rubber hoses – they fill up with oil – and it leaves the grit. Before refilling the machine with oil, pull the tank plate, wipe out the tank, then swap out the filters. Most people figure it's a new machine so why do they have to do that. But it's important."*

When times are slow, Langsdon suggests that's the best time to perform preventive maintenance on machinery, and use the employees productively. Still, he notes that for many people it takes a 'catastrophic event' before they pay attention to their machinery maintenance needs. *"The key word in all of this is preventative,"* he stresses. *"If you wait until the machine can't function, it's too late. Now you've gone from maintenance to a repair, and the cost is a lot more."*

Langsdon says that a molding machine in good repair operates at maximum efficiency and there are also some power savings to be realized. *"We had the Tennessee Valley Authority (TVA) come in and do an audit for us, and just in air leaks alone we were losing $8,000 a year. And that's just because of leaky air! It's the little things that get people and cost them a lot."* Compressed air leaks are also identified as a major cause of expense by cost and energy management expert Robin Kent, who says that typical leakage rates around a plant can be as high as 40%.[4]

KraussMaffei's Caprio advises processors to schedule planned preventive maintenance because that is more conducive to higher up-time on machinery, and fewer mid-production interruptions due to equipment going down. *"Our customers tend to be getting bigger with acquisitions and expansions, so up-time is critical to them. If they have the equipment OEM handle the calibration services and other PM services, they can often prevent unplanned major equipment failures. Clamp down time is key and most people understand that maintenance is key to achieving that."*

Another company providing training in mold maintenance, repair and related technology is ToolingDocs (**www.toolingdocs.com**). They provide certified training and also services such as Maintenance Capability Assessments. Their website also provides access to some free resources such as a Maintenance Calculator, which can be used in planning a maintenance schedule based on mold usage data.

MoldTrax5 from **www.moldtrax.com** is a customizable software documentation system which is claimed to reduce repair costs and unscheduled downtime, helping to increase productivity and free-up technical resources. The standardised documentation allows repeating problems to be identified and resolved for example, and also includes components such as troubleshooting guides and 'Tech Tips'.

12.5.1 Precise wear measurement helps to avoid unscheduled downtimes

Nothing drives processors crazier than unscheduled downtime on a machine. For example, worn barrels are a ticking time bomb threatening to affect product quality and cause severe gearbox damage, according to KraussMaffei Berstorff (**www.berstorff.com**). Regular wear measurement should be carried out on the

4 Kent, R.J., Cost Management in Plastics Processing: Strategies, targets, techniques and tools, **2007**, Plastics Information Direct, pp 148-9, 978-1-906479-00-8.

figure of eight bores of their twin-screw extruders to get detailed information on the barrel condition early on before real trouble starts.

The measuring device used by KraussMaffei Berstorff for the barrel housing bore detects wear in the processing section by measuring the inside diameter of each bore using an endoscopic camera on the measuring head. During the measuring process the actual diameter of the bore is measured every 5 mm over the entire processing section length and this procedure is carried out three times in different positions in each of the two bores. The result obtained is a representation of the inside diameter in the form of six tracks (3 left, 3 right) and with an axial resolution of 5 mm. The degree of wear is calculated by the evaluation software on the basis of the individual diameter values. In addition to production defects, wear measurement carried out at an early stage helps to avoid damage to the extruder.

KraussMaffei Berstorff's endoscopic measuring device also saves time. Previously it was necessary to disassemble the entire processing section during longer shutdown periods in order to precisely locate wear insider the extruder barrels. Today, disassembly is no longer necessary, and a complete wear check takes just a couple of hours. According to KraussMaffei Berstorff, many customers benefit from this service within the frame of preventive maintenance in order to avoid substantial gearbox damage, product quality defects or reduction of output rates.

Keeping molding machines and other equipment, well maintained and in good condition prevents unscheduled downtime, and improves both capacity utilization and profits.

12.6 Summary

Equipment is the heart of your business. The markets you are able to serve depend upon the type of equipment you have, and the way you use it. Your equipment can help you be profitable or can drain your bank account. Equipment selection is important, as is maintenance.

The molding press is the biggest investment, but the auxiliary equipment that you will need adds to the expense (e.g. chillers to cool the water that runs through the mold; desiccant material dryers to remove moisture from the resin so the materials will process optimally; hoppers to hold the resin as it feeds into the barrel (some come as a single unit called 'hopper-dryers') and grinders to regrind runners and other scrap materials/parts for reuse or for sale to other molders or resin brokers.

According to the type of products made, you will also require inspection and testing equipment such as comparators, measurement equipment, etc. In addition, increasing numbers of molders are offering added value with secondary-operations equipment as discussed above for assembly or decoration.

Technological advances in machinery and secondary operations equipment have definitely raised the bar for entrance into the custom molding business, but also offer opportunities for molding companies to be more than 'just molders'. Custom molding companies can be full-service contract manufacturers providing many additional services to the OEM, and increasing their own opportunities for profitability as a consequence. However, technological advancements have also leveled the playing field among competitors globally because anyone with access to money can purchase this technology. The differentiating factor then becomes how the technology is used to find creative solutions to the manufacturing challenges that OEMs encounter, and optimized to give the molder a good return on investment (ROI).

Chapter 13

Moldmaking technology

> ☞ **One thing is certain; no one can get a perfectly conforming part out of a mold that is not built correctly.**

Many molding company have in-house moldmaking capabilities. In fact, many molding companies got their start as moldmaking shops and evolved into molding as a way to add new business opportunities and to eliminate the ups and downs of the moldmaking business. Or, as one moldmaker told me, "*Moldmaking is a one-off business. You make a mold, get paid and that's it. Molding parts is like getting paid every day, 24/7.*" Still, there are also a lot of stand-alone mold companies that specialize in mold design and manufacturing. Many have expanded their capabilities into rapid prototyping, often now described more broadly and accurately as 'additive manufacturing' – to provide greater depth of engineering services on the front end. Many others have expanded capabilities at the other end of the process with mold qualification and tryouts, pre-production part runs or even building complete molding cells for their customers with molding press, automation (robots, conveyors and vision systems) and mold.

Moldmaking is critical to successful molding. The quality of any plastic part or component is typically 50% dependent upon the mold and 50% dependent upon the molder. The mold, after all, is one of the many variables in the molding process. One thing is certain; no one can get a perfectly conforming part out of a mold that is not built correctly. Molders and moldmakers have to make complex decisions about where to invest their money in this area in order to deliver cost-effective high-quality parts.

The moldmaking industry has undergone some tremendous changes during the first ten years of this new millennium, adopting new technology and working practices, but it has also been hurt by many factors including the push to have molds built in low-cost countries, the global downturn in the automotive industry, and the wider economic recession. According to the AMBA, the American Mold Builders Association (**www.amba.org**), approximately 10% of their membership went out of business during the past decade, and the organization actually lost 25% of their members just from bad economic conditions during the 2008-2009 recession. AMBA went from an all-time high of 450 members in 2000 to approximately 250 in 2010. The Canadian Association of Mouldmakers (**www.camm.ca**) has also suffered major losses as some very large mold manufacturers went out of business during this last recession, leading the Canadian government to provide stimulus funding in the worst hit area.

The automotive industry had a lot to do with the recent troubles among North American mold manufacturers. Their continued pressure on these critical suppliers to reduce prices and accept payment terms that could extend as long as two years, as well as the bankruptcy of many of the tier one suppliers put these companies into a bind that many couldn't withstand. Competition from Chinese mold companies also ate into much of the business of many shops, initially undercutting them by 50-60%. While many OEMs are beginning to see the pitfalls of having molds built in low-cost countries, there continues to be pressure put upon North American mold shops to offer a 'China solution.' However, the gap between North American and Chinese shops has also narrowed, some say to approximately 20-30%, which if the OEM adds in shipping and other costs associated with having to monitor a mold being built half-way around the world, results in a more level playing field than it was ten years ago.

However, the arrival of Chinese moldmakers on the scene has certainly forced the North American and European mold manufacturers to be much better at what they do. Back in the mid-1990s, I heard Bill Kushmaul, of US mold builder Tech Mold Inc. (**www.techmold.com**) say at an AMBA convention, "*We have to get the man-hours out of the mold if we want to be competitive*." That was when a 24-week lead time for a large, multi-cavity mold was still acceptable. Now, mold manufacturers have largely managed to get the man-hours out of the mold through various routes, including improved software and automated machine tool technology. The 24-week lead time is gone. Backlog at a mold shop is nearly unheard of, as lead times have dropped to as little as four weeks or even less.

13.1 CAD/CAM software

Case history: Mirplex Molds, Inc.

Development time continues to be a bottleneck for many mold manufacturers, particularly with the many changes that occur when working collaboratively with customers. A good example of a solution to reducing lead times is provided by Mirplex Molds, Inc., a 25-year-old Canadian mold manufacturing company (**www.mirplex.com**). Mirplex molds range from small, multi-cavity molds to very large molds up to 15 tons per half. Like many mold manufacturers, the company found itself getting bogged down in mold design and development. New CAD software from Siemens, the NX CAD/CAM digital product development system (**www.siemens.com/nx**), together with their

Sinumerik computer Numerical Control (CNC) technology, helped Mirplex reduce this time by enabling mold design and tool path creation to take place in parallel, because design changes from the customer can be incorporated more easily, reducing overall lead time.

Mirplex have estimated that mold design takes 25% less time with the NX systems, in part due to faster incorporation of customers' design changes, which now takes about 40% less time that it did previously. "*NX Mold Design has standardized our processes and we have a library of parts such as mold bases that we can reuse so that the mold is already half-designed when we start a project,*" a spokesman said.

Siemens has added more and more functionality in its NX CAD/CAM systems over time, including part validation. There are modules of the software that do part validation, checking wall thickness, curvature, or undercuts which you can apply to the model and get a fast readout. According to Siemens, molders often start with a less complex 2D solution, although many have already migrated to 3D, or are in the process of doing so. Good 3D design tools enable the moldmaker to work quickly on the mold geometry to accommodate customer modifications to part design.

In a White Paper produced by Parametric Technology Corp., Louis Gary Lamit from DeAnza College in Cupertino talked about the benefits of 3D vs. 2D technology. 2D can be cumbersome and difficult to use as a true design tool, and the only areas of similarity between these two methodologies are the use of 2D sketching for geometry creation and the availability of a plotted/printed document at the end of the design process. Documentation is easier to extract with a 3D CAD system, and the outputs are actually intelligent extensions of the total database. While they may contain drawing files, they can also offer a complex video-like quality. 3D also offers interconnectivity so that any changes to the documentation propagate to the component and assembly. Changes to the component propagate to the assembly and the drawings, and changes to the assembly propagate to the drawings and to the component.

The Siemens product is just one example, and various CAD/CAM software programs have proven ideal for mold designers and manufacturers, including Pro/E, AutoCAD/Autodesk, Delcam, Vero and others. Because mold designers work so closely with their customers, a good CAD/CAM program is essential to avoiding missing dimensions, changes to one program that don't get picked up by another, and other common mishaps that can cause expensive changes down the road. Many mold manufacturers have gone to 3D modeling for more realistic, more accurate imaging of the parts.

13.2 Manufacturing cells for moldmaking

Mold manufacturers began setting up manufacturing cells in the 1990s when automation came on the scene in a big way, allowing the use of robotics in equipment such as high-speed CNC machining centers with automated drill-bit changers for example. Some of this equipment had as many as 60 sizes of drill-bits. As machine tools became more highly automated, mold shops began using specialty machinists, and the job of the moldmaker changed from one of fully taking charge of a mold build from start to finish, to performing specialty tasks and overseeing several automated machining and electrical discharge machining (EDM) centers in a work cell. Thus, fewer actual moldmakers are needed, allowing shops to reduce staff. Using specialty machinists to do most of the machining work, the moldmakers are available for assembling the molds during the final phase of mold building.

Some cells have up to four machines in them, and may include CNC machining centers, an EDM center, and pallets that carry the mold components. Typically one man oversees the cell. **The moldmaking cell technology has reduced the number of moldmakers required and hence the number of man-hours in the mold build**, and helped North American and European shops become more competitive again.

13.3 High-speed machining (HSM)

One of the major advances in moldmaking technology over the first decade of the 21st century is high-speed machining. Most machine tool manufacturers have high-speed CNC machining centers with large, palletized tool changers that can make short work of cutting steel. Even milling of very hard steels can be easier with HSM. It's rare than a company won't have a high-speed machining center because this technology has reduced the man hours that it takes to build a mold, as well as the lead time. In fact this was probably the primary technology that helped European and US moldmakers compete most effectively with the Chinese shops in terms of price.

Learning to optimize the benefits of HSM hasn't been easy, requiring retraining of machinists comfortable with slower speeds, and the machine is only as good as the operator behind the controls. HSM has contributed to the reduction of lead times and backlogs, which moldmakers both like and dislike. They are keen to reduce lead times to meet the demands of OEM customers, but some still like the apparent security of backlogs. With mold times compressed from the 20-24 weeks of a decade ago, typically to 8-10 weeks today, backlog is nearly a thing of the past.

13.4 New technologies from moldmakers

Mold manufacturers have a lot of engineering expertise in-house, and a lot of experience in what moldmakers need to produce better molds for their customers. Often, they come up with creative and innovative ideas that they patent and sell commercially or license to the other mold manufacturers.

Case histories: Tech Mold Inc. and BA Die Mold Inc.

For example, Tech Mold Inc., a manufacturer of precision, multi-cavity, multi-technology injection molds for Fortune 500 customers in the medical, pharmaceutical and packaging markets, has made available its Universal Electrode Guide that ensures EDM small-hole precision with faster set-ups and diameter changes.

The Guide remains fixed in position on the machine, with its centerline precisely in line with the spindle, but it is easily adjustable to accommodate the broad range of small to large electrode diameters. This removes the need to maintain an inventory of dozens of different sized electrode guides, or to wait for a special-sized guide to be made and delivered, as well as time consuming inspection. Set-ups are quick and easy, and multiple hole sizes in a single part can be accomplished with predictable accuracy since the centerline of any electrode held in the Guide is always in a known location and is referenced by the machine's X & Y axes.

BA Die Mold Inc. (**www.badiemold.com**) received a patent on its Programmable Electric Rotating Cores (PERC) system that claims to revolutionize complex unscrewing cores used to make threaded parts such as caps and closures. As moldmakers know, conventional unscrewing molds are hydraulically driven, and the number and size of the threads on the part dictates the number of revolutions the core must rotate before the mold can open. The number of rotations necessary then determines the length of the hydraulic rack and the size of the hydraulic cylinder.

With a conventional unscrewing mold, gears would be used to step down the length of the rack. "*This works with many threaded parts but not all,*" says Alan Petrucci, founder of BA Die Mold. "*The gearing can take up a lot of real estate in the mold base, which becomes an issue and creates limitations when there are other actions necessary for other features on the part or if the processor needs to fit the mold into a particular size press.*"

The all-electric servo-controlled PERC System offers greater efficiency than a hydraulically system. For this reason, some of BA Die Mold's customers specify PERC even when the unscrewing can be effectively achieved with a conventional hydraulic rack and cylinder. Another benefit is that the PERC

System is fully programmable and extremely accurate, said Petrucci. "*You can program an unlimited number of profiles and repeatability is absolute. It's perfect for electric molding machines – the controls for the PERC System can be integrated/ hard-wired right into the press, much like a core pull function.*"

PERC eliminates the need for a rack using a servo motor to rotate the core. That means parts with an unlimited number of threads, intricate geometry and other special requirements are now possible and practical. Eliminating the hydraulic cylinder also brings the usual benefits in terms of cleanliness and maintenance.

13.5 Offshore mold manufacturing – to do it or not

Many mold manufacturers have attempted to provide what some of their OEM customers want and get a Chinese mold manufacturing source. They've done this with joint ventures, or by working with a broker who knows the sources, and will typically work with them to get the mold built, overseeing the build, ensuring its safe delivery, and ensuring that it makes conforming parts.

Having a mold built in China – the most-frequently targeted low-cost country – can be challenging. There are many pitfalls that can trap the average mold manufacturer, and result in a disastrous experience. So how can you ensure success if you decide that building a mold in China is necessary?

Kelly Wilson has spent his entire career in the plastics industry including managing custom molding operations, and overseeing molds being made in China. Now with Atlantis Industries Corp. (**www.atlantisusa.com**), he still gets molds made in China. Based on this experience Wilson says that if you want to be successful, you need to ask the following questions:

1. What is the real landed price to my factory? There can be extra costs involved, such as fees to bring the mold from the mainland to Hong Kong. Be sure you know what those costs are up front.
2. Who covers freight costs from China? Again, spell this out from the beginning. The tool might be cheap, but things can start adding up.
3. Is it understood that the purchase order is for a tool to produce ***good*** parts, not just samples, for a number of years? It is essential to watch the mold sampling, either in person or via video.

4. Was the tooling sampled in an equivalent press to the one in which it will run in your plant? You really want to know what machine you're going to run the mold in. A lot of these sample houses are just that – sample houses. They may be throwing your mold in a 400-ton press because that's the one that's open when you're going to run it in a 220-ton press. It doesn't have to be the same exact brand, but it needs to be same tonnage, pressures and shot size.
5. Did the tool run for 24 hours or just long enough to get samples? Not every mold needs to run for 24 hours, but it's a good test of how the mold operates. As the mold settles in things can change. A lot of time they'll sample molds without water, but water in China can be dirty. Also, request that they use resins that you supply for sampling. A lot of the materials in China you can't trust to be the equivalent material as in the US. Spend the money and send the material.
6. Did you receive parts on the runner and not groomed? That will tell you a lot.
7. Are mold components used available in the USA? That's becoming more and more common as the major mold component suppliers like Progressive and DME have established facilities in China. But, spell it out on the PO that the BOM (Bill of Materials) has to be approved prior to mold build. We want to be able to buy spares, and better yet, have them make your spares there.
8. Also, you bought the electrodes so you need those packed in with the tool, along with anything special used to make the tool. You need to ask for these to be shipped with the mold.
9. Does the purchase order include all CAD files required to make domestic changes to the mold? Make sure you have the entire CAD file. Also, when you go through and make changes did those get translated to the file? Everyone does this, not just China, where we've got the file but had the mold builder groom something to make it work and that didn't get reflected onto the file.
10. Are CAD files updated to reflect the actual tool being shipped? Does your CAD file actually reflect the part being made? What did you change in the last two weeks during sampling that means we don't know the latest dimension?
11. Will the mold be used in China or shipped to your own plant? You may be willing to accept a lower quality if the mold is remaining overseas.

Wilson has learnt how to work successfully with Chinese mold shops over 15 years, and stresses the value of going there in person, but also seeking advice from other companies with experience, and developing good working relationships in order to ensure you get the quality and service you expect.[1]

1 Goldsberry, C., Ten questions you must ask before buying a mold in China, March 5, **2010**, www.plasticstoday.com.

13.6 10 Steps to being competitive in moldmaking

1. Implement new technology such as high speed machining (HSM) to get the job done faster and more accurately.

2. Train your employees on an ongoing basis, particularly with respect to HSM, so that they can maximize what the machinery and equipment can do.

3. Automate as much as possible with robotic work cells so that the human factor is kept to those things that truly require the expertise and talent of the moldmaker.

4. Think outside the box: be more than just a moldmaker; be creative and innovative in finding solutions to customers manufacturing challenges.

5. Don't be afraid to present these new ideas to a customer. They might say, "Just make it to print," but they might also see the tremendous advantage to a new way of designing the mold that will give them reduced cost-to-manufacture or better cycle times or higher quality.

6. Go further downstream and further upstream in offering capabilities, including R&D assistance, making rapid prototypes or prototype or developmental tooling on the front end, and perform mold qualification or validation capabilities on the back end. Extend your reach and expand your capabilities.

7. Become true 'partners' with your OEM or molder customers, and get to know their business as intimately as you know your own, so that you can anticipate those things that they require.

8. Be collaborative with your customers throughout the mold design and build process, and keep them apprised of the progress. Customers don't like surprises.

9. Educate your customers about molds and mold building. Tell them why you quoted the mold you did, what are the benefits of the type of mold you quoted. Explain the costs involved, such as the expense of a hot runner system, and the ultimate benefits.

10. Market your company's capabilities and expertise, and don't undersell yourself. You know the value of what you provide, so sell that value and customers will be willing to pay your price.

13.7 Make or buy mold components?

Whether to make all your mold components or buy a proportion of them is a huge decision that every mold manufacturer must make in the quoting process because it will ultimately influence your competitiveness.

13.7.1 The moldmakers' experience

For many years, moldmakers made nearly all of the mold components that went into their molds, from mold bases to ejector pins. They were proud of the fact that they made the entire mold in-house, and pointed out to customers that this allowed them total control over the mold build and its quality. Today, most have realized the economies of scale that come with buying many components and even mold bases, from outside. One moldmaker asked: "*Why would I want to use my high-speed machining center to hog out a mold base?*"

Tech Mold's Jerry Seidelman has said that moldmakers need to evaluate the make or buy aspect of mold components, and consider the following:

- Consider efficiencies – buy more components on the outside. Seek the best decision on the make-or-buy. There are more and more component manufacturers – keep an eye on the value you can get from the outside and how that can help you to be more profitable.

- Evaluate your internal capabilities: Capital, equipment and people are all critical components in this business. Know what you can do, what you can do best, i.e. profitably, and then determine what jobs are money makers and which ones will be money losers. Reject the money losers.

- Provide more solutions for your customer – creative solutions. When you become a commodity, you're dead. Be a problem solver for your customers.

Franchino Mold & Engineering (**www.franchino.com**), specializes in very large, custom molds. Sales engineer Dave Clouse says for that reason, Franchino is limited in the off-the-shelf mold bases they can purchase. "*We used to buy a lot of DME mold bases but our market has migrated away from that size product,*" says Clouse. "*We do order custom plates from time to time if the scope of the program calls for it. There's no sense in fabricating a bunch of plates when we can buy a fabricated unit that makes perfect sense.*"

Clouse says that most of the mold components the company purchases are more affordable than they could make themselves. "*We buy a lot of Progressive components, such as side interlocks and core pins*" notes Clouse. "*If it's very* *specialized we'll build it in-house, but most tool shops aren't set up to do something like that as cost effectively as the component suppliers can do it. Side interlocks for example – male and female – run about $500. We couldn't do it for that price if we had $2,000 to spend on it. However, some parting line locks we'll fabricate, if they are very specialized."*

Like most mold manufacturers today, Clouse says that Franchino's job is to engineer the tool, engineer the proper components, and decide what they're going to purchase, with the goal of getting the fast lead time that the customer wants. *"The less time people have their hands on a mold the more cost effective it is. Making components can absolutely destroy the profit you have in a job. Machine utilization in a job shop is key. The more components you have to put on a mold the more stuff you have to worry about getting through a shop in a timely manner. If you decide to make all the components yourself, the amount of money you'll spend setting up to make a few components will really cut into your profit. These mold component companies make components day in and day out."*

Jonathan Winters, from component supplier PCS Company (**www.pcs-company.com**), agrees with that statement. *"It wouldn't make sense from a cost standpoint with so many of us making components at a fraction of what it would cost the mold shops to do it in-house."* He has also raised the issue of lead times. *"Certainly moldmakers are being driven by their customers with respect to lead time, they don't have time to buy the stock, set it up and make it themselves with the short lead times being demanded in today's market."* With respect to mold bases, Winters said that 10 years ago, standard mold bases represented about 75% of the company's orders, with modified bases accounting for only about 20-25% of their business. *"That's flip-flopped today,"* Winters said. *"We're now doing more modified bases. We have to turn these bases – with pockets, water lines, locks and pillars – in record time."*

Michael Mirante, owner of M&M Tooling Inc. (mmtooling.com) reported an 'uptick' in business at the beginning of 2010, with customers outsourcing their mold bases rather than making them in-house. *"That's why we're doing primarily mold bases now,"* Mirante says. *"The busier moldmakers get the more they tend to buy on the outside. When they're not so busy, some will choose to utilize their higher investment machines on mold base work just to keep the machines busy. Some have to be dead slow before they'll do a mold base in-house."*

Mirante explains that it's the core and cavity and detail work that is more financially rewarding, so they would rather use their high-speed machines for that. Additionally, he adds, *"a lot of these guys aren't tooled up for mold base work. You don't necessarily need high-speed machines but it goes both ways. Some are building mold bases in high-speed machine fashion if they have enough capacity and don't need their machining centers for core and cavity work. The machine has to be tooled up differently as well, and many don't want to be retooling their machines all the time to accommodate mold base work."*

The machining time is quick, Mirante adds, but the set-up, fixturing and programming takes time. *"Once it's tooled, programmed and set-up, the machine can run lights out"* he says. *"Moldmakers can't continually be tooling up back and forth, because the margins are so lean. We can't afford to tool up our machines all the time."*

A company such as Graphic Tool Corp. (graphictool.com) makes some of its own mold bases, and sources others outside to release capacity in-house. Graphic Tool specializes in high-cavitation medium to large molds up to 12,000 lbs. Rich Burman of Graphic recently purchased a new Mazak horizontal milling machine with pallets to replace two old Mazaks that were so old they were no longer holding tolerances. He was using the old Mazaks to rough out the steel for a base, then putting it in a Makino machining center to finish it. *"I like to build all my bases in-house because I can control the delivery and schedule the build the way I want,"* comments Burman. *"And I can control the quality, but I also use good quality mold base suppliers."*

13.7.2 The component supplier's viewpoint

Glenn Starkey, President of Progressive Components (**www.procomps.com**), a leading supplier of components for molds and dies, says that his company is working to standardize many of those items that started out as custom-made components to make it easy for mold manufacturers to buy ready-made mold components. *"We want to make it easy for the mold designer, efficient for the mold builder and convenient for the molder,"* Starkey says. *"The communication between the three is not always seamless. That's where we find that we need to get in and present the options to all three in order to do the math for a new component, to review together any concerns they might have."*

Progressive offers three levels of components: standard off-the-shelf components; special components, which are components such as a standard core pin with some detail or a collapsible core with threads; and custom components. *"Custom components require a print for us to make, such as a custom expandable cavity with some part detail in it or a custom lock,"* Starkey explains. *"We've optimized those processes so that the communication aspect is as expedient as possible. We're working to develop templates for custom items to eliminate the process of sending prints, and we have worked to then make not only custom components faster for quicker delivery – more like a special component – but also, over time, to standardize some of the special products so they can be purchased as off-the-shelf items. That's the Progressive model of the evolution of our components."*

The bottom line to this issue, according to Starkey, is "*Buy what you can and make only what you absolutely need to make, in-house.*"

13.8 Summary

1. **If you have in-house mold manufacturing that can be a value add for molders.** Many got their start in molding through their mold manufacturing businesses.

2. **If you don't have in-house mold manufacturing, there are plenty of mold manufacturing companies with which you can partner.** There is still a lot of capacity in the moldmaking business, so finding a partner that can build molds for your OEM customers isn't a problem.

3. If you have in-house moldmaking, carefully **evaluate what mold components you should make and which should you purchase.** Make the optimum use of your machinery and equipment on those parts of the mold for which you can add the most value.

4. If you decide to get molds built offshore, there are a lot of questions you need to ask before you go. **There are risks to any mold build, and those risks go up when you are having a mold built half-way around the world.** Evaluate the 'price' of the tool versus the 'cost' of the tool over the long term.

5. **Learning to maximize your mold manufacturing equipment is key to improving profitability in moldmaking.** The idea that you can lose money on the mold and make it up in molding the parts, is confounded thinking. Lost profits are lost profits, no matter how you cut it.

Chapter 14

Additive manufacturing – from prototyping to production

14.1 Evolution of the business

Additive or rapid manufacturing has itself developed rapidly over the last two decades, with great advances in technology, but also changes in the business, and most importantly, a fall in the cost. It began as rapid prototyping, and consisted of powdered UV-curable photopolymers and UV lasers used in a process known as '**stereolithography**' or SLA. Layers of material are built up, using laser light to solidify the material and make parts that are good for use as 'touchy-feely' samples for engineering and marketing departments to look at and determine if the component is feasible. The initial limitations of the rapid process included lack of actual thermoplastic materials, which meant the parts were definitely just 'samples' (no functional testing could be done with them), and the high cost of the equipment used to produce these samples.

However, developments came rather quickly. Next came '**selective laser sintering**' or SLS, which improved upon SLA to give better part quality. SLS uses a high-power laser to fuse particles of powder to create a part. **PolyJet** patented by Objet (**www.objet.com**) 'jets' curable polymers in ultra-thin layers (16 microns) with each layer cured with UV light after jetting, which creates parts that are ideal for concept models and master patterns. **Fused deposition modeling** (FDM) became popular and offered another option in the 'rapid' processes using a plastic filament of material and an extrusion nozzle to extrude in layers. FDM notably delivers parts in the actual plastic materials such as ABS and polycarbonate, and even high-performance engineering grades such as Ultem polyetherimide, that can be used for functional testing and low-volume manufacturing – this represents a big step forward from having to use alternative materials.

In the early days there were business problems in the RP industry. Service bureaus became the model for having RP parts produced, and while many sprang up throughout the US and Europe, in the US they developed problems. While the technology was fascinating, and seen as an interrupter to older, slower technologies such as machining wooden samples or using patterning, for all their investment RP just didn't catch on in a big enough way to make these bureaus profitable. Gradually, during the 1990s, most of the large-scale service bureaus went out of business.

While there was a lull in that business model, RP equipment and material companies continued down the path of development. 3D Systems Corporation (**www.3dsystems.com**), considered a leading innovator of rapid prototyping, was quickly moving into **3D printing**, another form of additive manufacturing in which a three dimensional part is created layer-by-layer, and other 'rapid' solutions. 3D Systems has been instrumental in recent years in bringing rapid prototyping from the enclave of the service bureau to the desktop of any design engineer that wants to create a part.

A number of other companies were also developing new processes, materials and equipment to compete with 3D Systems. Those included Electro Optical Systems (EOS) in Germany (**www.eos.info**), today one of the most advanced developers of new technologies such as **direct metal laser sintering** (DMLS), that can create end-use parts from various metals that have the strength of machined metal parts. This has proved to be the disruptive technology to machined components in many applications, including injection mold cores and cavities.

The service bureau model didn't completely die, as many OEMs or product designers/developers find it easier and more cost effective to use the AM process to get a few parts rather than get a prototype mold (typically one cavity) made. (However, there can be benefits to a prototype mold as well.) Today there are many service bureaus that provide a variety of 'rapid' services for an increasingly interested clientele. These service bureaus include companies such as Quickparts.com which is now a part of 3D Systems, Morris Technologies (**www.morristech.com**), Harvest Technologies (**www.harvesttech.com**), Proto3000 3D Engineering Solutions (**www.proto3000.com**), Vista Technologies (**www.vistatek.com**), Laser Prototypes Europe Ltd. (**www.laserproto.com**), CA Models (**www.camodels.co.uk**), Digital Mechanics (**www.digitalmechanics.com**), Schneider Prototyping (**www.schneider-prototyping.de**) and Solid Concepts Inc. (**www.solidconcepts.com**) which does direct digital manufacturing including SLS, LSA and HDSL (high-definition stereolithography) in its five facilities in the US, among others. These service bureaus work with mold manufacturers and molders whose customers request a prototype part for evaluation by engineering or marketing. Part files are uploaded to the service bureau via an FTP site and the part files are then downloaded to the equipment for the AM processing.

The additive manufacturing (AM) industry has become stronger over its 22-year history (in spite of some ups and downs), and this newer name provides a better descriptor of what the various processes can actually do, as opposed to the original name of '**rapid prototyping**.' Still, demand remains for actual prototyping processes, notes Scott Harmon, VP of Business Development for Z Corporation (**www.zcorp.com**). *"We think in general the use of printed RP parts for form, fit and function is reasonably well understood by design engineers, and we also see growth in the classic RP industry applications."* (As opposed to additive manufacturing of functional parts from powdered polymer materials such as ABS, PC etc.) The company is currently experiencing year-on-year growth.

Terry Wohlers of industry experts Wohlers Associates (**www.wohlersassociates.com**) would agree with that assessment, based on his latest analysis published in Spring 2011. The compound annual growth rate (CAGR) of revenues across the additive manufacturing sector (equipment, materials and services) in 2010 was 24.1%, and double digit growth is anticipated for several years to come.

In the last couple of years additive manufacturing has really started to move into end product manufacturing, with two key market areas driving this development: highly specialised, often short-run or one-off and typically **high-value components**, e,g, for the medical, dental, aerospace, automotive, and motorsports industries, and **personalization** of consumer products e.g. in furniture and home accessories, toys, game avatars and fashion products.

Two main trends have also been seen in the development of additive manufacturing equipment with new systems brought to market recently by a number of manufacturers, including EOS (**www.eos.info**), Objet Geometries (**www.objet.com**), MCOR Technologies (**www.mcortechnologies.com**), Optomec (**www.optomec.com**), 3D Systems (**www.3dsystems.com**) and Z Corporation (**www.zcorp.com**). These are either larger equipment with a larger build area for parts, or smaller 'desktop' printers that can make small prototype parts quickly. Some of these new desktop printers cost less than $5,000, making them available to many engineering offices.

☞ **Double digit growth is anticipated for several years to come.**

14.1.1 New service bureau models

A newer model for an additive manufacturing service provider is targeted clearly at designers and consumers, with one example being Netherlands-based Shapeways (**www.shapeways.com**). The company, which launched in 2008, allows customers to upload a design that Shapeways will manufacture using a variety of AM techniques. Unlike many other providers however, Shapeways focuses entirely on the consumer market, and also offers a portfolio of 'creator' tools that makes it easier for customers unfamiliar with conventional design tools to create custom products. In addition Shapeways offer artists and designers a facility to set up their own online 'storefront' to sell AM-produced products to the public, where Shapeways handle the sales transaction, manufacturing and shipment to the customer.

Quickparts.com in the US is also one of the most notable of the current additive manufacturing service providers. Ron Hollis founded Quickparts.com in 1999, soon after the collapse of many of the large service bureaus that provided rapid prototyping services during the 1990s. Quickparts.com quickly rose to the top of the new generation of service bureaus offering rapid prototyping (SLA, SLS, FDM, Polyjet), cast urethane parts, injection molding (tooling and parts), sheet metal parts, CNC parts and plaster and metal casting. Today, Quickparts.com is recognized by many manufacturers, including those in the plastics industry, as one of the best in the additive manufacturing business. Hollis is also author of a book on additive manufacturing: Better Be Running! Tools to Drive Design Success.[1]

14.2 Latest materials and systems

In 2008, chemicals giant Huntsman Advanced Materials (**www.huntsman.com/advanced_materials**) announced the development of an entirely new additive manufacturing process and machine. Based on entirely new MEMS (**micro-electro mechanical system**) technology, Araldite Digitalis is fundamentally different to other technologies, and is neither based on lasers nor on light reflecting MEMS as used in 3D printers. At the heart of Araldite Digitalis is an MLS MicroLightSwitch – a radical new exposure system operating via a computer-controlled micro-mechanical shutter system, which selectively exposes a larger surface area of radiation curable resin in a single step. Compared with a laser which cures one point at a time, this larger exposure area enables much faster and accurate manufacturing, even of very complex parts.

Objet Geometries' (**www.objet.com**) latest Connex 3D printing systems are the first systems in the world to support **multi-material 3D printing**, according to Objet. The systems offer product developers the ability to print parts and assemblies made of several materials with different mechanical and physical properties in a single build process. They use patented PolyJet Matrix Technology, which jets materials in ultra-thin layers, immediately curing each layer with UV light, resulting in models that closely emulate the look, feel and function of an exceptionally wide variety of end products.

In 2009 Z Corporation altered the competitive landscape by introducing the world's first automated, **monochrome 3D printer**, which was the company's most

1 Hollis, R.L., Better Be Running! Tools to Drive Design Success, **2007**, CLSi, 978-0979576027.

affordable, automated 3D printer to date. Even with its time-saving automation features to simplify printing, office-safe build materials, aggressive dust-control, and zero liquid waste, the new machine creates finished parts at a fraction of the cost and effort of other systems. According to Joe Titlow, Director of Product Management, speaking at the time *"The ZPrinter 350 is an ideal choice for organizations seeking the most affordable way to produce single color prototypes fast for early concept evaluation and functional testing, painted parts for a finished look, or patterns for casting applications"*.

The vertical build speed is 0.8"/hour (20mm/hour), in a full size build box, 200 x 250 x 200mm. *"A build box of this size allows designers and engineers more flexibility in part creation,"* said Titlow. *"They can do larger-sized full parts rather than partial parts that must be adhered together. It also gives people the ability to build several smaller parts at once."*

Material for Z Corporation's RP printers is a plaster-based engineering composite powder in a binder solution. Justifying this choice of material, Titlow said *"We look at the materials from a slightly different perspective. Even for people designing plastic parts, the up-front phase is primarily initial concepts, so we're very focused on the front end where all the innovation happens. The better you are at doing the front end, the more innovative your products will be. To be able to do lots of concepts is the key, even though it might not be the exact material the product will be in, but that's not important at the front end."*

Z Corporations's materials can be treated in various ways so engineers can test the function of a particular design. *"There are no fatigue or ductility properties to our material, but RP parts can be functionally evaluated,"* Titlow explained. *"A part made in an injection mold tool, at the end of the day, will be very different than most RP parts. You can't RP a perfect surrogate. The key is testing the design until you tool up. Primarily, RP for Z Corp. is getting perfect simulation, and what you value up front is the ability to generate a part with accuracy and detail."* The ZPrinter 350 also provides for integrated recycling of unused material and a self-monitoring feature.

DSM Desotech (**www.dsm.com**), introduced its latest advancement in materials at Rapid 2010, the annual additive manufacturing trade show run by the Society of Manufacturing Engineers. Somos NeXt, is claimed by DSM to be the industry's most advanced resin today for approaching true thermoplastic performance. According to the company, Somos NeXt produces parts that have a combination of stiffness and toughness which typically characterizes thermoplastics, yet with all of the high feature detail, dimensional accuracy and aesthetics offered by stereolithography.

"This is a material you need to hold in your hand to fully appreciate," says Vince Adams, DSM Somos marketing manager. *"The stiffness and toughness combination*

produces a look and feel that is so like a thermoplastic, people are really surprised that it's actually stereolithography." Parts built from Somos NeXt actually perform like engineered thermoplastic, which provides customers with the ability to do functional testing, according to Don Portenga, sales engineer at Eagle Design & Technology, a rapid product development company whose capabilities include SLA, SLS and urethane casting.

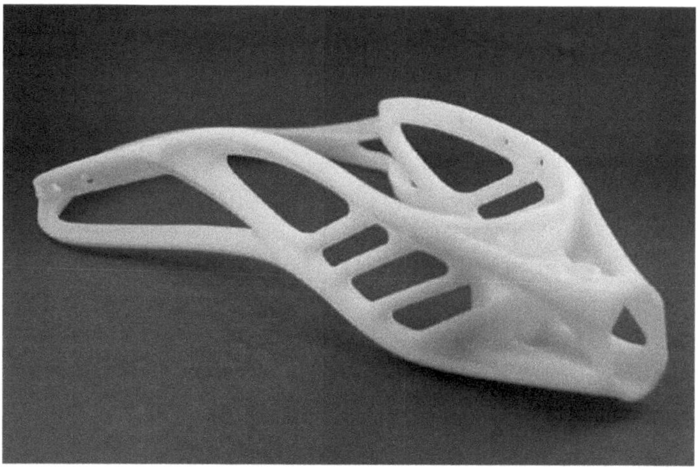

Figure 14.1

Somos NeXt was used to create this Lacrosse prototype for Warrior Sports *(DSM Functional Materials)*

However, while many materials have the goal of the 'injection molded part' look and feel, and even pushing the envelope for functional testing, there continues to be demand for the true prototype. With equipment price points coming down and more desk-top 'printers' available, this is becoming more possible at the engineer level. Z Corp. claim that there's still lots of room in the 'concept modeling' arena. *"When you read the literature and talk to advanced design organizations, they use concept models to get in front of customers, product research organizations, and test groups, and there's a lot of value to that. They get to engage the customer with the design and interact with them on a different level than with just a computer screen image or sketches and written descriptions. Really smart firms are pushing the nature of that interaction to a physical level with Z Corp parts. We can make parts that look molded with full color so the customer can get the full sense of what the product or part really looks and feels like. Growth in both segments of this market – AM and rapid prototypes – will help achieve future technologies in this industry."*

14.2.1 Virtual prototyping

Virtual or digital prototyping such as in the Autodesk Inventor system essentially means 'part-less' 3D prototyping, enabling complex products to be 'built' on-screen.

As additive manufacturing has become more accessible in price and possible build size has increased, so the case for virtual prototyping has been harder to make for many applications. Most product designers, developers and marketers still want a part they can hold in their hand and in some cases, put into an actual product to test its functionality. Nonetheless, accurate on-screen 3D modelling has been shown to help companies design better products, reduce development costs and bring products to market faster. Designs can be shared, evaluated and refined more effectively, without the time and expense of creating physical prototypes at every stage of the process.

14.2.2 Direct metal laser sintering

Direct metal laser sintering (DMLS) has entered the picture slowly, but can definitely be seen as a disruptive technology to machined metal parts, including cores and cavities for injection molds. EOS (**www.eos.info**) has become a leader in the manufacture of laser sintering systems, and in the development of new materials. The company recently introduced several new metal materials for its EOSINT M 270 systems including EOS NickelAlloy IN718 and EOS Aluminum AISi10Mg.

EOS NickelAlloy IN718 is a nickel-based heat resistant superalloy which corresponds to the commonly-known Inconel 718 alloy, which means the material can be used in applications that require high temperatures. Greg Morris, president of pilot-user Morris Technologies Inc. (**www.morristech.com**), said that company had been supplying parts in EOS CobaltChrome MP1 as a preferred material for high-stress/high-temperature environments since 2005. *"Since working with EOS NickelAlloy IN718, this material has become our front-line material of choice for many applications."*

Morris also provides cores/cavities for injection molding applications that the company claims can sometimes yield as many as 50,000 parts, depending upon the material used. This would be boon for many moldmakers and molders, since the DMLS process can produce these cores/cavities in just a few days.

14.3 Moldmakers and molders get in on the action

With more options in new equipment and materials than ever before, more companies are making additive manufacturing a part of their molding and/or moldmaking operations. Companies using the technology to their competitive advantage are finding that additive manufacturing can help them expand their services, particularly in R&D, prototyping and short run production parts, and interest in additive manufacturing is booming.

Case histories: Molders offer additive manufacturing

One example of how molders are using additive manufacturing or rapid prototyping to their competitive advantage is provided by Action Mold (**www.actionmold.com**), a custom injection moldmaker and molder with 12 molding presses. Their Action Prototype & Urethane Casting division offers a one-stop shop for fast turn-around R&D prototypes and production parts.

In an effort to provide better service to its customers in the area of product development, Action partnered with Chris Wentworth, an experienced industry professional who has more than 15 years experience running successful prototype and cast urethane operations. "*Many of the low-volume parts don't go to injection molding initially, and in the past we've had to send these customers elsewhere to get prototypes or low volumes parts, then wait for them to come back to Action for the high volume production,*" said Wentworth. "*A lot of product designers, if they don't go to a shop that does both prototyping and moldmaking, often end up with something that isn't moldable. Going to someone who does both makes sure it's manufacturable, so that if down the road they need 100,000 parts, they know from the start that it's moldable.*"

Action Prototype & Urethane Casting and Action Mold together can take a project from concept to production. "*Together we provide a partnership that can assist in product design, development, prototype and production all under one roof,*" says Wentworth. "*That's a tremendous competitive edge in today's market.*"

Mack Molding's Application Development Center and Mack Prototype, Inc. (**www.mackprototype.com**), reportedly run two Stratasys FDM machines to enhance Mack's engineering capabilities for its medical OEMs, again adding a lot of value to its custom molding operations. Each machine has unique capabilities to fit the needs of the two divisions. Randy Pell, Sr. Staff Design Engineer in the Application Development Center, notes that the FDM machine in his division is smaller and builds parts in ABS resin in an 8" x 8" x 10" (200 x 200 x 250mm) build envelop, and builds in 0.010" (0.25mm) thick layers. "*We use the FDM equipment to perform design verification and to ensure that snap-fit parts are functional, as just a couple of examples,*" he says. "*Using this allows us to set up our quality processes, such as how to measure a part, determine the fixtures we'll need, and even print out our fixtures using the FDM machine. In the orthopedic market, many of the components used in surgery might look identical, but actually differ in size by a few millimeters, so we can set up fixtures* *to ensure each component is being measured correctly. We can help our customers bring new products online faster."*

Mack Prototype Inc. has an FDM machine (Fortus 400) that runs two different materials – ABS 30i (ISO 10993 biocompatible material), and a PC-ABS – which allows the division to make parts that can not only be used in testing and clinical trials, but can actually be an end-use product. The machine will print a much larger part – 14" x 16" x 14" (360 x 400 x 360mm) in resolutions between 0.005" (0.13mm) and 0.013" (0.33mm) thick layers – which is one of the largest build envelopes available through a custom manufacturer in the New England region, notes Ric Perry, president of Mack Prototype. *"We're finding that we do a lot of CNC machining of components and this FDM process is closer to a machined part than the SLA was,"* says Perry. *"We can make parts in ABS or PC-ABS, and in a parallel path with the machined parts. We can be machining multiple parts as well. The FDM process is also suitable for low volume parts. It can be cost effective to print the parts rather than making a metal mold and molding parts, depending on the part and the geometry."*

Pell added that end-use resins are strong enough today to work in surgical environments, where metal used to be the only suitable material, and thus reduce costs. *"Hospitals have to autoclave metal instruments; so by offering a way to convert a metal instrument into resin, we can help our customers bring down the cost of the instrument by making the actual instruments disposable. In this instance, we use FDM to print out resins that would be like the disposable trials, so you wouldn't have to build prototype molds, another time and cost savings."*

Morris Technologies (**www.morristech.com**) in Cincinnati, Ohio, is another company that has taken full advantage of AM technology over the past decade. The company is heavily involved in the aerospace, medical and automotive markets. Morris has an SLA laboratory that is fully equipped with six 3D SLA machines, including one running DSM's NanoTool high-temperature resin machine, and a Connex500 from Objet Technologies that is capable of running two resins at a time with different properties.

The company's DMLS laboratory is reported to be outfitted with equipment from EOS, including one EOS 250Xtended machine and five EOS 270s, which can take parts from 3D CAD files to production quality parts in hours. Morris specializes in building DMLS 'rapid tools' (RT) as Morris calls them, and injection molding parts from these at its medical business unit. The company also has SLS (selected laser sintering) capabilities. The company operates a total of six DMLS systems at their main facility and another three at Rapid Quality Manufacturing (RQM), a sister company that specializes in the production of additive metal parts using DMLS.

"DMLS might seem expensive until you take into consideration the time you save," says Mark Combs, inside sales manager for Morris. *"It's actually very cost effective when you need fast time-to-market. Some things you can't do conventionally, plus it gives you a lot of design options because part geometry isn't a limiting factor. DMLS opens up so many possibilities."*

Titanium AM is the company's newest capability. This material runs in an M270Xtended machine equipped with a micro-filtration system in an argon gas environment. Morris say that there are many applications for building parts in titanium, especially in the medical and aerospace industries.

14.4 Making additive manufacturing a part of your business

Almost every industry can benefit from AM, as this process has become an important part of the engineering product development cycle. This means that if you as a molder and/or moldmaker can provide this service you have increased your value to your customer base. Molders and moldmakers don't necessarily need to purchase these AM machines for their own facilities, especially since there are today many reputable service bureaus throughout North America and Europe. Partnering with these companies and offering their 'rapid' or additive technologies to your customers is a good way to get in on the action of AM without a lot of capital expenditure.

The service bureaus offer FTP sites onto which you can upload your part print files and they can create your part and have it shipped in less than 24 hours. This is a good relationship to have as it provides an additional service for your customers at little expense. Many parts, depending upon size and complexity, can be made quickly for under $2,000. To many design engineers this is a bargain, as it means that design can be finalized and the production mold started in a timely manner. Time-to-market is important to OEMs.

Some molders have purchased 3D printers, FDM or SLA machines to use in their mold engineering departments, and it has paid off in terms of being able to do the upfront engineering, then building the production mold and doing the molding. Many molders and moldmakers complain that their OEM customers don't come to them early enough in the game for them to provide effective, efficient production molding solutions. Adding AM to the mix of services you provide may encourage customers to come to you early and work with you on product development.

However, this will only add profitability if you remember that these services cost money – whether you go outside and partner with a service bureau or

bring equipment in-house. That means you'll have to charge for these early, upfront 'rapid' engineering services. However, since most engineers now see the value in being able to produce 'prototype' parts quickly, and the ease of creating many iterations in a short period of time, the cost is generally a minimal factor.

Case history: Industrial Molds Group, how one moldmaker makes AM part of its business model[2]

Customers at Industrial Molds Group (industrialmolds.com), a 60-person mold manufacturing company in Rockford, IL, often asked about 'prototype' technology and how they can get prototyped parts.

Tim Peterson, Vice President of Industrial Molds Group, says, "*There are two kinds of prototyping, to our way of thinking. There is what is called 'rapid' prototyping, which can be done on 3D printers in a very quick way when you just need a part to see what it looks like, touch it, have marketing examine it for consumer acceptability, and get a feel for what it might be when put into production. We can provide this by helping with the part design, then coordinating the outsourcing of the FDM or SLA parts that are required for a visualization or marketing evaluation. We have contacts with several prototyping service companies we work with locally.*"

The second way of looking at prototypes involves making actual parts that are dimensionally correct from the exact material required for the finished product. Peterson has to make sure that the customer understands the difference so that Industrial Molds Group can provide the customers with what they actually need. "*What we typically do when we are asked to provide these types of parts is what we call a 'Limited Production Prototype' (LPP) mold,*" Peterson explains. "*What this gives our customers is a production quality part in a near production or production setting.*"

There is a cost differential between get a rapid prototyped part and a part from an LPP process, in which Industrial Molds Group provides a single cavity mold with the core and cavity cut to exact dimensions. "*While there are some benefits to getting just a part in SLA or FDM, or even a part from a desktop 3D printer, we believe there is real value to building an LPP mold,*" says Peterson. "*One of the primary benefits is that we take all the information we learned from the R&D on the LPP mold, and replicate that in a multi-cavity production tool.*"

A company specializing in prototyping can sometimes make a core and cavity conventionally, and put in a MUD (multi-unit die) base that they own to

2 Goldsberry C., reprinted by permission of Industrial Molds Group from their July 2011 company e-newsletter.

provide the parts to the customer. After the parts are molded however, if the company wants the mold, they only get cavity and core, not the entire mold.

The suitability of prototype parts for a project also depends on the complexity of the mold. *"If you have a complex mold, you might have to use hand-loaded inserts, and that may not be representative of a production mold,"* noted Peterson. *"Let's face it, you won't be making a million parts using a hand-load tool. Somebody sees a rapid prototype part and thinks it can be done, but maybe not in a production molding setting."*

That's valuable information for Industrial Molds and for its customers. *"With rapid prototyped parts, you haven't really learned anything about the mold or the part in the actual production setting,"* says Peterson. *"Most of our customers are automotive with highly engineered components. Often they want actual parts that can be put on a vehicle and tested in real time. We've worked very hard to perfect our production side, but we're on distinctly different paths from prototype parts."*

Through its LPP molds, Industrial Molds Group's customers learn a lot about the part, optimum design and the manufacturability by building a mold that provides 'pre-production' or 'limited production' parts. After R&D and collaborative engineering on the LPP mold, that mold can even be used as a 'bridge' tool to supply parts until the multi-cavity mold is completed.

"The key to understanding what prototypes can do for your project including providing valuable marketing information, optimizing mold design and production capabilities and improving time-to-market, is in knowing what you're trying to accomplish," Peterson advises. *"Do you just need a quick 3D part to see what it will look like? Or do you need something more?"*

There's some intrinsic value to the FDM or SLA or SLS, but the true value is really determined by the customer. *"For example, if we're working on a cosmetic part with high aesthetics we can hand-work this component, get a few samples so our customer can get it to marketing to see if the product will be acceptable by the consumer, before getting a production mold. This helps eliminate a bad product on the market,"* said Peterson.

Typically the value of the prototype part is driven off the fact that it goes into production, and the prototype helps you get there. But if you're not sure the product is going to production, then don't spend the money on a production mold.

For any prototyping requirement, the biggest value is in understanding the objective. *"If you just want to look at a part, feel the part, and need 10 parts for marketing and engineering to evaluate, you don't want a pre-production tool,"* says Peterson. *"If you need production quality part that's dimensionally correct to put on some vehicles and do testing to make sure it will work – a production quality part – then it's more cost effective to invest in an LPP. But it's important to know which is which and what you really want."*

Chapter 15

Polymer materials – minimize costs and maximize choice

For molders, who are essentially buying in plastics materials and reshaping them into products, raw materials costs constitute a major proportion of their total production expenditure. According to industry expert in cost management, Robin Kent, in the case of mass produced goods, materials can account for as much as 80% of total costs, and even for technical products 45% is typical.[1] It is not difficult to see that making small savings in materials expenditure can be crucial to the success or survival of a molding business.

> ☞ **In the case of mass produced goods, materials can account for as much as 80% of total costs, and even for technical products 45% is typical.**

15.1 The price of plastics

The price of raw materials has become a huge variable at molding companies, and one that changes almost on a monthly basis. Many years ago, it was common to buy materials on a contract basis, particularly those materials that the molder bought in large quantities, in which the price was fixed for a certain period of time. During the first decade of this century, material prices have fluctuated wildly, rising rapidly, and coming down more slowly if at all. Resin suppliers therefore have been more reluctant to give contract pricing even for their biggest and best customers. Smaller custom molding companies are often forced to buy materials on the spot market, where they can buy a few hundred pounds on the spot and take delivery quickly. This typically costs more.

Trying to track these cost increases for raw materials, and then pass them along to customers, is difficult at best. Customers don't like it when the price of the parts fluctuates from one month to the next, but trying to second-guess where the resin pricing will be from one month to the next is nearly impossible. Managing inventory becomes extremely important in a business climate such as this.

1 Kent, R.J., Cost Management in Plastics Processing: Strategies, Targets, Techniques and Tools, **2007**, Plastics Information Direct, 978-1-906479-00-8.

'Over-buying' material can seem like a good idea when prices are on the rise, or when more specialised materials are in short supply, but you must consider the real implications on cashflow and plant operation. Can you really afford to tie up money and devote factory space to storage? Poorly stored materials can cost more money in the long run, potentially requiring more drying before processing, and even getting lost or spilt if stored inappropriately. You may even find yourself left with unused material if a customer changes his specification or removes his order entirely.

In many cases, the types of material a custom molder uses depends upon the markets served. Some markets require high-end thermoplastic materials that are quite costly such as polyetherimide and polyamide-imide, high-temperature, chemically-resistant materials used in aerospace and automotive engines, or highly specialised medical grades for bioresorbable implants for example. Expertise in molding specialist materials can differentiate your company from the competition, but you need to make sure you are quoting realistically for jobs which may require higher molding temperatures or additional compounding for example, which can increase energy consumption or tie up plant and personnel for longer.

15.1.1 The price of scrap

In light of the increasing material costs, molders are becoming more aware of issues such as scrap, and making efforts to reduce the amount of scrap generated. Hot runner system have seen increased demand to help with this issue, as well as new purging compounds that require less material for changeovers or color changes. Even commodity materials that were once considered relative cheap in price are now expensive, so wasted material is wasted money.

Reducing materials wastage is a far more significant issue than often realized, and one that can be easily addressed. According to Robin Kent, when companies carry out a simple waste assessment they can often find waste costs 20 times higher than they initially estimated. For a firm making a gross margin of 7%, a reduction in waste costs of just 1% has the same effect as increasing sales by 14%! **Wasting materials, together with the man hours and services consumed in their processing, is literally pouring money down the drain, and giving away a real competitive advantage.**

However, reducing scrap is only one part of the cost efficiency equation. Understanding more about the materials you use can also help you to minimise production costs, and also to quote realistically for jobs. **Sometimes you can work with a customer to specify the best combination of material and part design, resulting in the lightest, strongest part to meet his requirements.** Your client may require ever-higher service temperatures, or he may be keen to explore greener options. The more you know about new materials, including additives and processing aids, the more effective you can be in recommending the best materials, and also minimising your own production costs. So make sure you stay well informed about the latest grades. Sign up to email news services and industry magazines – your customers will value the knowledge you can share.

15.2 Material developments

New materials are often developed for specific new applications, either as replacement materials for metal components, or replacing other types of thermoplastics to provide enhanced properties. Two or more decades ago, resin producers developed new resins, and then sold them into various industries for upcoming applications. Nowadays there is less speculative development, as resin producers are more careful to design new grades for specific applications. Nonetheless there are still some impressive advances which are helping plastics to replace metals and other materials, reduce product costs and improve performance in service, and meet legislative requirements. Keeping abreast of new materials and grades makes sure you offer the best possible service to customers.

Building good relationships with compounders can also be worthwhile. Compounders will very often work with molders on a new material for a specific application, so just because a customer comes to you with a unique product that requires some unique material properties doesn't mean you have to turn them away. Compounders can help determine the material that is required, which additives or fillers might be required and blend a specialty compound that will work for the application.

15.2.1 Automotive materials

For example, Ticona's new Fortron 6162XF is a mineral/glass-reinforced polyphenylene sulfide (PPS) with superior fuel resistance. Ticona is a global leader in polymers for fuel system applications, and developed Fortron 6162XF to meet changing industry requirements for higher operating temperatures in highly oxygenated, more aggressive fuels. This new low-swell fuel-resistant Fortron PPS grade reduced weight gain by more than 45% compared with Fortron 6165A6, their existing grade which had also been used successfully in fuel systems. In 1,000-hour exposure tests in CM15A fuel conducted at 194F (90C), Fortron 6162XF exhibited superior fuel resistance, with a weight change of less than 1 percent.

Both Fortron PPS grades are well suited for applications that need extremely high dimensional stability during prolonged exposure to aggressive fuels. Applications can include impellers, outlet covers and fuel injector components. As a high-performance material, Fortron PPS meets the challenges posed by today's automotive fuels system environments, specifically excellent creep resistance even at elevated temperatures, virtually no moisture absorption (less than or equal to 0.02 percent) and potential weight savings of up to 50 percent versus metal, leading to greater fuel economy (**www.ticona.com**).

The automotive industry has always represented one of the biggest markets for polymers. Now that industry is developing new hybrid, energy-efficient vehicles where both weight and environmental friendliness matters, material suppliers are working to meet those requirements. DSM Engineering Plastics recently introduced a new grade of its Stanyl nylon material for under-the-bonnet applications in the automotive industry. Stanyl Diablo OCD2300 and also Akulon Diablo (nylon 6) are new products which the company claims provide cost optimization and productivity gains, long term sustainability and reduced environmental impact (**www.dsm.com**).

Requirements for under-the-bonnet applications are continually evolving as environmental requirements, European EURO V and VI legislation concerning emissions and the call for reduced fuel consumption have brought changes such as the use of smaller engines with higher turbo pressures and EGE (Exhaust Gas Recirculation). Consequently components such as air ducts, air intake manifolds and charge air cooler end-caps are now exposed to higher operating temperatures.

DSM claim that Stanyl Diablo OCD2300 and its predecessor Stanyl Diablo OCD2100 outperform existing high heat resins on high temperature stability, but also on weldability, weld strength and the weld's long term heat stability, giving guaranteed part reliability for the lifetime required. Akulon Diablo increases metal replacement options for under-the-bonnet applications, and is claimed to meet all the needs of the supply chain for air ducts.

15.2.2 Compounded materials for electrical and electronic applications[2]

Electrical and electronic products are a significant market for compounded materials. Applications include computers and office equipment, telecommunications equipment, electrical and electronic machinery, electronic components, instruments and related products. Good opportunities are expected

2 Goldsberry C., Market Snapshot – Electrical & Electronic, April, **2010**, www.plasticstoday.com.

for compounded thermoplastics based on a continuing need for lightweight, high strength, and heat and corrosion resistant materials as obsolete equipment is replaced. Electrical conductivity or insulating behaviour is frequently important.

Opportunities in mobile phones, personal digital assistants (PDAs), games consoles and other consumer electronics will be based on the increasing importance of aesthetics and design features as manufacturers strive to differentiate their products in a competitive environment. Scratch, impact and fingerprint resistance can all be important in these frequently handled devices. The role of compounders in this segment will continue to increase alongside rising demand for highly customized components and the subsequent close interaction between material supplier and end-product manufacturer. Leading additives used in electrical and electronic products include stabilizers, colorants, flame retardants and conductive materials such as carbon black.

RTP Company (**www.rtpcompany.com**) a major compounder based in Winona, MN, worked with one of its customers, Connectors Unlimited (**www.connectorsunlimited.com**), a producer of interconnect solutions for the electrical and electronics industry, to formulate and supply plastic materials for a shielding solution for analog cable and circuit connectors. Several static dissipative and conductive compounds were chosen for the improved static control they provided. "*The fundamental reason for using RTP Company's conductive plastics in our analog connectors is to minimize noise interference in the electronic signal, which is primarily caused by static charge,*" said Marty Ignasiak, President of Connectors Unlimited. "*By effectively reducing static charges, we're able to significantly improve the signal-to-noise ratio in our connectors and thus upgrade overall performance of devices.*"

Base resins for the compounds are RTP 300 series polycarbonate, and RTP 3400 series liquid crystal polymer (LCP). To maintain rigidity during assembly (possibly including soldering processes), glass and carbon fiber reinforcement are utilized along with flame retardant packages to meet UL94 flammability requirements. The LCP materials are claimed to provide an invaluable option when components must undergo electronic or solvent cleaning.

In 2009 Ticona Engineering Polymers added a new low-chlorine Vectra liquid crystal polymer (LCP) alloy with PPS to its portfolio of V-0 (flame retardant) halogen-free polymers. The new LCP material has improved processability and equivalent properties to flame-retarded nylon grades, and gives designers and manufacturers a 'drop-in' option for use in eco-friendly connectors and electrical components. According to Ticona, V143LC is a high performance inherently flame retardant (FR) thermoplastic without flame retardant additives, which can withstand high soldering temperatures used for surface mount technology in the electrical and electronics industry, including those needed with lead-free solder (short-term up to 340°C).

15.2.3 Materials in the personal care market[3]

According to PolyOne, thermoplastic elastomers (TPEs) have revolutionized the personal care industry. TPEs are found in a multitude of personal care products including toothbrushes, hair brushes, razors and cosmetic packaging, giving a designer the opportunity to differentiate their product performance and aesthetics from the competition. Very often, two material molding is used to produce these personal care items. A substrate of ABS, PC, or other material is molded, then a TPE is overmolded onto the substrate to give consumers a nice feel, a better grip (useful in the case of razors, for example where hands may be wet), and colorful looks for branding purposes (**www.glstpes.com**).

PolyOne claim that their GLS TPE materials offer longer life and improved durability. They also provide impact and chemical resistance, excellent aesthetics and color retention, and improved tactile feel. While personal care spending globally is down, the growth of TPEs in this market has exceeded the overall market growth.

15.3 'Green' materials

As more companies want to provide consumers with 'green' products, many resin producers and compounders are developing materials that use post-industrial and post-consumer recycled materials; as well as producing materials from bio-materials such as plant materials, fibers, starches, and other bio-based products to make plastics compostable and degradable. Molders are finding that customers often ask about these materials for their products so that they can achieve the 'green' qualities they look for to help promote their products.

This is a confusing area, with many new materials on the market claiming green credentials, but one that can help molders maintain a competitive advantage. Materials suppliers can provide information about processing, performance in service and ultimate environmental impact (e.g. life cycle analysis) all of which may be more or less important depending on the application in which the product will be used. Relative cost compared with 'traditional' plastics is also an important consideration for most customers. Interest in all things green will certainly continue

---

3 Goldsberry C., Market Snapshot – Personal Care, December, **2009,** www.plasticstoday.com.

to grow in the coming years so it is important to keep abreast of new developments. Magazines such as Injection World (**www.injectionworld.com**) and Compounding World (**www.compoundingworld.com**) run regular features on these materials.

According to The Freedonia Group, global demand for bioplastics, those plastic resins that are biodegradable or derived from plant-based sources, is expected rise more than fourfold to 900,000 metric tons in 2013, valued at $2.6 billion. Growth will be fueled by a number of factors, including consumer demand for more environmentally-sustainable products, the development of bio-based feedstocks for commodity plastic resins, and increasing restrictions on the use of non-degradable plastic products, particularly plastic bags. Most important will be the expected continuation of high crude oil and natural gas prices, which will allow bioplastics to become more cost competitive with petroleum-based resins.[4]

Biodegradable plastics such as starch-based resins, polylactic acid (PLA) and degradable polyesters, have accounted for the greatest proportion of bioplastics demand until now. Double-digit gains are expected to continue going forward for bioplastics, fueled in part by the emergence on the commercial market of polyhydroxy-alkanoates (PHAs). PLA will also see strong advances in demand as new production capacity comes online. **However, in the next few years, non-biodegradable plant-based plastics are likely to be the primary driver of bioplastics demand**, posting extraordinary growth from a small base. Both Braskem (**www.braskem.com**) and Dow Chemical (**www.dow.com**) plan to open plants in Brazil in 2012 that will produce polyethylene from sugar cane-based ethanol, while Solvay is expected to open a bio-based polyvinyl chloride facility, also by 2012 (**www.solvayplastics.com**).

Western Europe was the largest regional market for bioplastics in 2008, accounting for about 40 percent of world demand. Bioplastics sales in the region benefit from strong consumer demand for biodegradable and plant-based products, a regulatory environment that favors bioplastics over petroleum resins, and an extensive infrastructure for composting. Going forward, however, demand will grow more rapidly in the Asia/Pacific region, which is forecast to surpass the Western European market by 2013. Gains will be stimulated by strong demand in Japan, which has focused internally on the replacement of petroleum-based plastics. Other regions such as Latin America and Eastern Europe will also see stellar gains.

PolyOne's GLS thermoplastic elastomer materials also come in 'green' compounds, such as the company's reSound biopolymer compounds and Versaflex BIO, both of which were recently introduced to the market. The reSound platform is formulated with a 30% minimum bio-derived content, and combines

4 World Bioplastics to 2013, Study #: 2548, November. **2009**, www.freedoniagroup.com.

compatible engineering thermoplastic resins with bio-based polymers such as PLA, polyhydroxybutyrate (PHB), polyhydroxybutyrate-valerate (PHBV), and bio-polyesters. Key improvements include high heat resistance of up to 120°C (248°F), and impact resistance up to 12 ft-lbs/in, according to PolyOne.

Versaflex BIO TPEs for injection molding are formulated with up to 70% renewable resources, compared with typical compounds that use a maximum of 15 or 20%. This product line features translucent grades, available in hardnesses ranging from 40 to 70 Shore A, that compare favorably with typical styrenic-based TPEs. They are expected to help designers address current and future environmental standards, such as the European regulations encompassed by REACH (**R**egistration, **E**valuation, **A**uthorisation and Restriction of **Ch**emical Substances), RoHS (**R**estriction **o**f the use of certain **H**azardous **S**ubstances), ELV(**E**nd-of-**L**ife **V**ehicles) and WEEE (**W**aste **E**lectrical and **E**lectronic **E**quipment).

RTP Company has developed a new series of **specialty compounds using post-consumer content** (PCC) resins. The reclaimed resins are plastics diverted from solid waste streams through recycling programs, and provide an additional environmentally conscious option to the material selection process. These PCC compounds can be tailored to target specific performance requirements while including properties from the company's portfolio of specialty compounds and thus provide a more sustainable, multi-solution product.

The percentage of reclaimed material in the polymer component of PCC compounds can range from one to 100% depending on the requirements of individual applications. The PCC compounds provide color, conductive, flame retardant, structural or wear resistant properties in addition to being easily customizable to meet corporate environmental sustainability goals or marketing objectives to increase consumer appeal.

RTP's PCC compounds are available globally in nylon (PA), polycarbonate (PC), and polyethylene terephthalate (PET) resins, as well as alloys based on these post-consumer feedstocks. The company claims that this range offers the versatility to be used across consumer goods, building materials, automotive components and electronics devices.

15.3.1 The downside of bioplastics

Bioplastics certainly don't yet offer a perfect solution to all the problems associated with the use of traditional oil-based plastics. There are genuine questions whether horticultural land can realistically be spared from food production to accommodate growing chemical feedstocks, but the main concern in the shorter term is the potential impact on the current recycling stream if bioplastics start to be included. Currently, there are well-established consumer and municipal recycling programs for certain materials, particularly HDPE and PET. These recycled feedstocks go to a variety of companies that require large amounts of recycled PET and other recycled materials to make products such as carpeting and fleece clothing; decking and other outdoor products. HDPE in particular can be re-used in the same product as before – namely milk bottles.

The fear is that the waste stream can inadvertently become contaminated by bioplastics, which don't process like traditional plastic materials, and these companies that depend upon recycled materials won't be able to make their products. There has been some work on the potential for reusing mixed PET and PLA, but the US National Association for PET Container Resources (NAPCOR) has refuted the premise that containers molded from PLA can be successfully mixed into the existing stream of recycled PET containers. They cite concerns over cost of separation, increased contamination and yield loss, and impact on recycled PET (RPET) quality and processing. While there might be technology in the future that would allow this, NAPCOR officials believe there are currently too many issues to overcome (**www.napcor.com**).

The Association of Postconsumer Plastic Recyclers (APR, **www.plasticsrecycling.org**) and NAPCOR have also made a joint statement to address what it sees as a growing problem resulting from the introduction of 'degradable additives' into the plastic packaging marketplace, and the negative impact these materials are likely to have on post-consumer plastics recycling, causing a serious disruption to the recycled materials supply chain. As recycled materials are increasingly used in higher-value products, degradable additives mixed in with recycled feedstock have the potential to contaminate the final product. APR and NAPCOR are both proposing that advocates for the degradable additives approach should have to prove 'no harm' both for first and subsequent recycled applications.

15.3.2 What this means to molders

What all of this means to the general custom molding industry is that bio-based resins will typically have to be handled differently from traditional resins, and probably handled separately to prevent contaminating the recycled resin supply chain. While the idea of bio-based resins has some validity in some products, these products currently are niche products that have special applications. Compostable single-use food packaging is an obvious and logical example. Molders that want to offer their customers the option of bio-based resins will need to learn to run these materials, which generally run at much lower temperatures than traditional resins.

Molders should certainly learn about these resins so that they in turn, can educate their customers about them. These options will be viable for some consumer packaging applications for example, but molders should not expect these materials to become wide-spread in their usage, particularly for durable products. Customers will certainly look to their molded parts suppliers to help them make these choices when it comes to their desire to manufacture products that are 'green' and thus more desirable to a segment of the consumer population. But until we completely understand the implications of these materials both in end-use applications and the recycling stream, they will have limited applications.

Final thoughts

Chapter 16

Planning for the longer term

16.1 What is the future of my business? Preparing to move on

Every business owner at some point begins thinking about life after his or her business, whether this involves a new venture or retirement. What will I do with the business if I have no partners or heirs that want to be involved? Sell it as a going concern? Close it down and sell the equipment? If you own the building, will you keep it to lease to another company? This can provide rental income during your retirement, and real estate is almost always a good longer term investment. The important thing is to plan for this stage in the life of your business as much as you would for any other. By planning ahead you can aim to sell up at the best possible price, and at a time that suits you.

16.1.1 Selling your business as a going concern

If you have followed all the advice in the earlier chapters of this book, then you should have an attractive and effective business to offer buyers, which should realize a good price.

If you decide to sell your business as a going concern what is your business worth? How do you evaluate it to come up with a good selling price? Knowing what your company is really worth is an important factor in deciding whether or when to sell up. Will selling now provide sufficient return for whatever you plan to do next in your life? Most company owners naturally believe their businesses are worth millions. They think back to all the years of effort they put into building the business, buying equipment, hiring good employees, building relationships with many good, long term customers, and suddenly in their minds their business is worth $50 million – even if it only does $5 million in annual sales. So, how much *is* your business worth?

To help answer that question, Patrick McNally of Blackman Kallick helps small-to-mid-sized companies evaluate their businesses and arrive at a selling price. In the May 2010 Blackman Kallick publication *Manufacturing Edge,* McNally says that businesses often miscalculate the value of their businesses. *"Rules of thumb, such as EBITDA (earnings before interest, taxes, depreciation, and amortization) and revenue multiples, can give inaccurate estimates of the value of your business, which can lead to bad decisions and missed opportunities,"* said McNally. *"Other business owners assume their balance sheet shows the value of their business. However, the value of most businesses is different from the balance sheet amounts, which are based on historical amounts, not current values."*

Much of a company's value might not be reflected on the balance sheet, such as fully depreciated equipment (a zero value on the balance sheet), or intangible values such as goodwill and customer relationships. The true value of a business, said McNally, *"is based on the cash flows that business can be expected to generate. To understand – and increase – value, you must understand what drives cash flow and what makes your business unique."*

- What do you do best and for whom?
- What can you offer that a purchaser may not find elsewhere?

However, before offering your business for sale you may also need to consider the following issues:

- Are you presenting your business in the best possible light?
- Does anything need fixing before you sell to get the best possible price?

To clarify these points, McNally advises that you ask yourself the following questions when considering selling your business and fixing a price. When you have completed this exercise you should be a lot clearer on the value of the business, and in a better position to fix a price and secure the best sale:

1. **What makes your customers buy?** Or, perhaps a better question is what makes your customers buy from *you?* Are your important customers buying for reasons beyond dollars and cents (such as a long-standing business relationship and confidence in your quality and delivery)? You may have a unique capability they require that is tough to find with just any 'me too' molder. These things are worth something!

2. **What is the value of the company's intellectual property?** Many molders might not think they have intellectual property because they don't have proprietary products per se. Rather, they make other manufacturers' products or product components. Yet, many custom molders and moldmakers do have intellectual property. Perhaps you've developed your own business or quoting software, or even some special mold design software. That's your intellectual property and it's worth something. Perhaps you've developed automated molding cells incorporating equipment that you manufactured

yourself. Some custom molding firms have even developed proprietary consumer products that they make, market and sell themselves. All of these add value.

3. **What about other hidden assets?** Do you have assets that would be difficult, risky, or time-consuming for someone else to duplicate? Does your business have information that other companies want, such as market intelligence or customer lists? Customer lists including contact information for the individuals that actually purchase molds and molded parts are extremely valuable in the industry.

4. **Is there a solid management team in place?** Having a solid non-ownership management team in place can be a huge boost to a company's value. Typically, buyers are wary of a business that is too heavily dependent upon a single owner - if you leave, will customers bail too? Ideally you need several people in your organization that have touch points with the customers, to enable you to retire but provide continuity for your successor.

5. **How well is the company managed, and is it utilizing its assets to its greatest advantage?** McNally has suggested that owners compare a variety of factors to those of other companies in the same industry, such as sales per dollar of fixed assets or employee, gross margin, and working capital turnover. If the business has a unique product or process, is it properly utilized or could it be more fully exploited?

6. **Does your balance sheet reflect the true cash flow and value of your company?** If you are taking disproportionate amounts of money out of the business, perhaps to fund family expenses for example, then the genuine performance of the company may not be accurately demonstrated by the figures.

7. **Are there opportunities to improve the top line?** Are appropriate resources devoted to sales and marketing, or are the sales people simple order takers? While that is less likely to be the case in the custom molding business, often marketing and sales is a badly neglected area, which is why we spent a lot of time and pages on it in the front of the book. That component of your business can do a lot to improve the top line, and boost the bottom line as well. Opportunities for sales growth can have a huge impact on the value of the business.

8. **How can operations be improved?** Can cash flow be improved by streamlining operations and cutting unnecessary costs? Are there non-essential assets that can be liquidated?

9. **Who might be interested in buying?** There are always potential buyers out there, and each one is looking for something a bit different than the other. And, they each have a different 'perception of value' when it comes to purchasing your company. To one customer your business might present an opportunity to move into a new product area, perhaps even exploiting a proprietary product or technology, whereas another might value your physical location close to an important customer base. Understanding who the buyers are and what is important to them is key to understanding the value of your business to those buyers.

16.1.2 When is the right time to sell?

Once you've completed an evaluation of your company's capabilities, and determined who you should be pitching it to, and what you should expect them to pay for it, you might be wondering just when is the right time to sell? There is no 'right' time which applies to every business. It's different for every owner. First you must ask yourself why you want to sell your business. According to Patrick McNally, People don't always sell for the right reasons. Some people sell because someone tells them the market may never be as good as it is right now. While that may be true, if you love what you do and have no compelling reasons to sell, you may choose to keep doing what you love even it means missing the top of the market.

McNally offered some important questions to ask yourself when you begin thinking about selling your company. But you don't want to wait until you're ready to retire right now before planning. You should ideally begin the process 10 years out by asking yourself the following types of questions:

- When would I like to retire?
- What will I do if I no longer have an office to go to?
- How much is my identity tied to the business?
- Do I define myself as a business owner or in some other way?
- How will the sale of the business affect my family?
- Do I have children who want to participate in the business and do they have the necessary skills and desire to keep the company running?

And, if the answer to that last question is yes then you next need to ask yourself:

- Do I trust that my children have the skills and management expertise to keep the business going? Do I trust them with my retirement income?

Once you have made the decision that you are ready to sell for the right reasons, you need to evaluate whether or not the business is actually ready for sale. This is critical to obtaining the maximum price for your company. Look at your company through the buyer's eyes. One way to do that is to bring a disinterested third party into your facility, to tell you from their point of view, how it really looks. Your perception isn't everyone's perception.

- While your equipment is in fine working order, and continues to do a good job, how old is your equipment? What does it look like?

- Look at the aesthetics of your facility. Is the production floor clean and neat, or is everything covered in a layer of oil and machining fluids that are 20 years old? Many of today's manufacturing facilities are beautiful, and the equipment and floors immaculate.

First impressions are important, and if your facility and equipment looks old and dirty then you're not going to maximize its value no matter how great your customer base is.

> ☞ **You don't want to wait until you're ready to retire right now before planning.**

16.1.3 Partners and partnerships

The question of how and when to sell up can be more complex when a company has multiple owners. Buy-sell agreements are critical if the company consists of a partnership between two equal owners or between several owners or family members. When there comes a time when one owner wants to get out of the business, or wants to sell his/her shares, a buy-sell agreement will prevent a lot of potential problems. In fact, a buy-sell agreement should always be part of the business plan between partners. It provides a route to buy out partners who no long want to be part of the company due to unforeseen circumstances, or for partners to sell.

In fact, it cannot be stressed enough that all businesses need to have plans in place that will secure the future of the partners and the employees should something happen to the partners. Many thriving, growing companies have been forced out of business due to a lack of proper planning for the future of the business.

16.1.4 Employee stock ownership plans

If you have a going concern and no obvious successor interested in taking over the day-to-day operations, one alternative to finding a new owner for your business might involve shared ownership with your workforce. These schemes are known as Employee Stock Ownership Plans or ESOPs in the US, and they offer one way for owners to gradually retire over a longer period of time. Candidate companies must be profitable and typically have dedicated long-term employees that have actively participated in running the company for many years. In the US ESOPs are well regulated and set up as trusts, overseen by an attorney. A good scheme rewards loyal employees by giving them an increasing share in a successful business, as well as benefitting the business by ensuring that employees have a greater personal interest in that success. In many ESOPs, employees have voting rights alongside management.

ESOPs are ideal for many small-to-mid-sized privately held companies, but there are some examples of larger firms. One such company is Nypro Inc., (**www.nypro.com**) the Massachusetts-based custom molder which was one of the largest companies ever to choose the ESOP route, in 1998.

Case history: Dynamic Tool & Design

Dynamic Tool and Design (**www.dyntool.com**) was founded in 1976 by John Lau and Irv Fisher, and with a top-notch management team grew into a strong company with a reputation for providing quality molds for its customers, many of whom have long term relationships with Dynamic. According to Irv Fisher "*Our focus has always been to grow the business, to build a lasting company that customers can count on to be there for them in the future. In addition we wanted the managers and employees who have worked so hard to ensure Dynamic's success to receive the benefits of this success.*"

Dynamic chose the ESOP route as the best means of ensuring this continuity for customers and colleagues. At the time of the ESOP average employee tenure was between 15 and 20 years, and all the employees who met the participation criteria became part of the ESOP.

16.2 Summary

The future of your business is largely in your hands. While you cannot control the whole industry or the global economy, you can control your business. You can make plans, you can create strategies, and you can prepare for the future by making sure you are doing everything you can to create a good, profitable business today. That doesn't happen by accident. If you think you'll want to sell your business in the future, you need to begin today to create a valuable business that someone will want to purchase and will bring in good money to you or your heirs. To wait until customers go out of business before designing a marketing and sales strategy, or to put off creating a business plan because you think you're not a large enough company to need one, or to delay making the proper legal plans to ensure that your business survives you means that your business has no future.

The future of your injection molding business begins today!

Additional material

Chapter 17

A case study

This case study demonstrates how the questionnaires in Chapter 2 can be used to develop a marketing strategy. It is based on an actual process carried out within a genuine moldmaking company, which also offers molding through a sister company. Names have been changed so the company and its clients cannot be identified.

The process is not complete, but already it can be seen that the company has lots of opportunities for improvement, and more importantly, the necessary information to start putting a strategy in place. Some of the problems identified are specific to this company, particularly their relationship with one major customer, however, this is certainly not a unique situation in the injection molding business. Moreover, there are many more factors highlighted by this exercise which will be familiar to lots of company owners reading this book.

Reading through this example may help you identify similar strengths and weaknesses in your own business, and also provide some thoughts as to how you can exploit the opportunities and minimise the threats. If you are asking a third party to question your employees, the responses here may provide some useful 'prompts' to elicit more information.

You can see from this example that the creation and implementation of a marketing strategy cannot be achieved instantly. It requires the buy-in of management and the will to drive through changes. However it can also be highly effective in holding up a mirror to your company, and seeing precisely where those changes can be made to most effect.

The first part shows the responses of employees at 'ABC Moldmaking' to the questions in Appendix 2.1, together with some initial comments and suggested actions. Part 2 offers some more detailed recommendations based on the data collected in Part 1. Part 3 then explains the parts of the process still to be undertaken using SWOT analyses and round table discussions, in order to complete the picture and define the marketing strategy going forward.

Part 1a: Who are we?

Typical responses provided by senior staff at a real moldmaking company are given below, along with comments and suggested actions based on these responses

Who are we? Provide your definition of the company.

- A high-quality, plastics mold manufacturer, and excel at automotive parts that have complex mechanicals.
- A very high-end, high-tech mold builder.
- A very high-end toolmaker with significant engineering capabilities.
- An injection molder and moldmaker for the automotive industry, also getting into medical products and consumer packaging, diversifying and specializing in industries that use complex tooling.
- A medium-to-large tool shop with a lot of capacity, expertise in tool manufacturing and technology, and providing solutions to our customers.
- We build challenging molds, and one area that excites us is tools with creative mechanics – out-of-the-box kind of stuff.
- We offer the ability for our customers to make changes later in the game, and are committed to building a good tool and back it to the fullest.
- One of the best, if not the best, moldmakers for automotive parts outside Chicago, that can build complex tools. We're a good option for customers looking for a full-service company that provide sampling services, PPAP, pre-production parts runs.
- A large tool shop that is very good at certain things, such as complex heavily engineered molds, with expertise in automotive.
- A plastic injection moldmaking company that builds all injection tooling, from single-cavity to high-end tooling using top-of-the-line technology and engineering.

Comments and suggested actions

The answers received to the question "Who are we?" were fairly consistent and show that management generally has a common understanding of ABC Moldmaking, its capabilities and its place in the market.

From these definitions, we can extract some words and phrases that might help develop articles, promotional materials, and case studies to be used to help get new work. Some of the descriptive words that sound good and define ABC Moldmaking strongly include:

high-quality; out-of-the-box solutions; mechanically complex molds; full-service, mold sampling/qualification (at sister company); PPAP, pre-production runs; significant engineering capabilities; creative mechanics; complex tooling; top-of-the-line technology and heavily engineered molds

Promotional materials can play on these words to develop the ABC Moldmaking Group brand and entrench the company as a company that builds "high-quality, mechanically complex, heavily engineered molds providing out-of-the-box, creative solutions using top-of-the-line technology, and full-service sampling, mold qualification, PPAP, and pre-production runs for customers with requirements in the engine and fuel components, fluid management systems, and other industrial industries, as well as packaging and consumer products markets."

What do we do (i.e. What type of capabilities, services, etc. do we provide to our customers?) – List 3 or 4 of our primary capabilities.

Build tight-tolerance tooling, Running lights-out overnight with robotics 23/7, High-speed machining, Unlimited mold building capabilities, Engineering services, Large tooling packages, Quality and precision, Customer service and account management, Product development, Craftsmanship in mold building, Automation, Full-service with Pyramid for sampling, Efficiencies – shortened lead times, Proactive in improving manufacturing abilities, Looking at solutions, Experience and energy (youth in our shop), Responsive in engineering and repair, Highly complex molds, Mold repairs, Multi-cavity tooling – high engineering added-value, Everything from design to PPAP, Troubleshooting, EDM and carbon cutting with robotic cells, Problem solving, Technical expertise, Capacity.

Comments and suggested actions

Obviously, ABC Moldmaking has a lot of capabilities and services. The key is to promote these capabilities and services, then maximize these so that you get a higher utilization in the shop. There is no doubt that ABC Moldmaking can provide extensive capabilities, services, experience and expertise in all areas of mold design and manufacturing, mold qualification, preproduction runs and beyond (through sister molding company).

It appears from this description that the company has a lot to sell to its customers and potential customers.

What do we do best, i.e. what are our greatest strengths? – List 3.

A. Challenging molds
B. Customer service
C. Tenacity

A. Sales
B. Experience on the shop floor (25 yrs. +)
C. Youth – 4-5 very skilled young people currently

A. Highly complex molds – up to 32-cavity
B. High-cycle molds built to last
C. Molds that cost less over the life of the program

A. Complex tooling with multiple actions
B. Very responsive to customers' lead times
C. Engineering

A. Mechanics - excel at creating molds that make unmoldable parts
B. Listen to customers- look ahead to keep customers from walking into problems
C. Never walk away from a commitment to a customer

A. Our people
B. Our systems – AQP or ISO systems
C. Tim's ability to sell us as a company which helped our customer base to diversify

A. Progressive thinking company-wide
B. Product development
C. Back-end services

A. Engineering
B. Product development
C. Commitment to the business

A. Big de-gas tools – bottles with ribs, etc.
B. Multi-cavity work

Comments and suggested actions

Based on these evaluations of ABC Moldmaking's greatest strengths, we can conclude that the three key areas to be promoted are (in bold):

***Challenging, highly complex molds** that **require engineering, design and product development expertise** as well as the **utilization of machine tool technology, experience of moldmakers, and commitment** on the part of ABC Moldmaking to its customers.*

What do we need to do better, i.e. what are our weaknesses? – List 3.

A. Simple molds (tend to over-engineer the simple things so we're not competitive on that stuff)
B. Inflexibility of the system (we have a system and we follow the system)
C. Engineering speed (we're built for speed at the back end but slow up front)

A. Grow the engineering staff (design and development)
B. Building – need to improve our facility for efficiency

A. Engineering is good but slow (problems with late deliveries)
B. Inherent problems with engineering – focus on customer service and make it painless for the buyer
C. Software limitations

A. Sense of urgency to meet deadlines
B. Customer service (e.g. late quotes)

A. More consistency needed in planning and organization processes
B. Repair work interferes with production, and more technology required

A. Streamline the flow of information – would require a change of culture
B. Feel the customer's pain
C. Push to improve – never be complacent

A. Efficiencies
B. Learn to use software as efficiently as possible
C. Technology utilization

A. Enhance engineering department – under-staffed, under-trained and use outside mold design help
B. HSM programming - a training issue
C. Job management – lack of leadership hinders completion of jobs

A. Need to be faster
B. Communications with customers needs to be better
C. Management/cultural issues as we grow. Can current management handle growth?

A. Communication is key
B. Planning from beginning to end – make a plan and stick to it

----

Comments and suggested actions

Most of the people interviewed seem to believe engineering is understaffed, which is causing a breakdown at the front end of the process and resulting in delays at the back end. While there does seem to be a system in place – and one that is rather inflexible – there is a problem with getting a project off the ground and flowing correctly.

Comments: "Hiring more engineers won't necessarily solve our problems. There's no driver - no engineering manager; no one willing to step in and tackle issues. Powering a project through engineering is a problem."

The company should explore whether this is purely a manpower problem or whether it is due to a lack of an engineering manager who can push projects through the front end.

To improve communications, you might consider looking at a plant management software program to provide a single point of contact between engineering, customer service and the production floor, and help with the flow.

On a scale of 1 to 10, with 10 being excellent, where would you rate our company in the following areas:

Overall quality of mold design:	8
Overall quality of our engineering capabilities/services:	8
Overall quality of molds:	8.3
Customer service:	8
Teamwork:	8
Communication – internal: 7, external:	8.5

Comments and suggested actions

Most respondents rated ABC Moldmaking's primary services quite high (mold design, engineering, molds built and customer service). Teamwork and internal communication were in the good range but generally people's comments didn't match the scores they gave. This shows a tendency to rate the company highly in spite of some of the problems they see.

Comments made in this section again referred to bottlenecks in Engineering and delays in quoting. There may be a problem with volumes of work, requiring another engineer, but the company should also look at systems and organisation. For example, are all engineers involved in quoting? Would this be better if just one or two people were responsible?

How would you compare our company with competing companies that you've read or heard about?

6 respondents rated ABC Moldmaking in the top 10% of the best mold companies.

3 respondents rated ABC Moldmaking in the top 20% of the best mold companies.

Part 1b: Who are our customers?

Describe the 'ideal' customer for our company? (Give 3 or 4 attributes of that customer)

- Responsive to our questions so we can meet their deadlines.
- One with high-end software – we have Pro/E and use Iges files to translate.
- A non-automotive customer. Maybe health-care/medical, consumer products.
- A medical, closure or consumer products customer whose molds utilize our automation.
- A customer wanting 1-4 cavity automotive molds, using high-speed machining and giving us high profitability.
- One that needs good tooling support (product issues, design needs), and no commodity tools as those types don't utilize our capacity.
- The customer that purchases $2 million-$8 million in tooling each year; with some knowledge of tooling, within a 200-mile radius, who understands the importance of using a top quality shop.
- Lower cavitation molds, looking for the expertise we have here, and allows us to help develop the program.
- One with products matching our experience – a lot of medium-sized tooling. We don't do well on very small or very large tooling. We struggle with through-put, put in massive amounts of overtime due to unrealistic expectations. We very rarely make money on very large tools. Automotive is good, but they have a ceiling on their price - we need another industry with similar products. We've done health-care molds very well, but those aren't very profitable.
- A customer that has a common vision with us – driven by value, not price.

Comments and suggested actions

Respondents generally have a clear view of what the company is good at, but the mixed responses suggest they may have less knowledge about which jobs are really profitable. Most wish to see diversification in customer base, but with molds similar to those that are currently built for the automotive industry.

More information should be shared about how each project really goes, whether or not the job went over on hours, why it went over, whether it made money or not, etc.

Name three of our best customers and one reason you rate them as such.

All respondents identified the company's two biggest customers – highlighting a likely overdependence (particularly as they are both in the automotive sector).

Comments and suggested actions

Respondents appear to understand why these customers continue to place orders with the company, and their respective importance in terms of the volume and value of their jobs. Some responses highlighted the difficulty in making a profit on lower budget tooling required by one of these customers.

Finding similar work, ideally in different market sectors, should be a priority.

In spite of appearing to know what the company does well, and the type of customer that requires this, marketing efforts to date have been fragmented, and have not brought in the desired new work.

What does our company need to do better to increase business and grow the company?

- We have to stay on top of the technology side like we're doing, but we should get into 5-axis high-speed machining. I think we're losing a bet there.
- Structured marketing – targeting industries and players that will utilize our equipment and engineering.
- Better communications up front and more automation in the back.
- More sales, better education in building more intricate molds efficiently. We're set up to do it but can't do it efficiently.

- More engineering help. We're an engineering-driven company and we have to make our engineering group strong – engineering and account management are tethered.
- Defining and understanding our business better, so we know what we are doing and why.
- Exploiting email marketing and developing an online store for mold components etc.
- Look at business areas beyond automotive; get it in the door and do it first. Too dependent on one company – need to diversify.
- Sales. We need more top sales people – dedicated to keeping up with high-tech software, high-speed machining (HSM) and robotics.
- Increase engineering speed, shorten the time line.

What capabilities do we need to add to achieve our growth objectives with current, satisfied customers?

- High-tech software, more HSM and robotics.
- Add to the head count.
- More engineers, but they're hard to find so we're understaffed.
- Repairs department that uses its expertise to be a money-maker.
- United sales front and more sales management.
- Continuing our education on HSM – better training on the technical products we put into our tools. More knowledge about molding.
- Education with respect to tool design, and also processing of materials.
- Need more automation, and to find younger staff to be with the company over the long haul.
- Communications – getting the right information from the customer at the right time.
- Marketing.
- 5-axis machining capabilities.

Comments and suggested actions

There is a perception that more equipment is needed, however there seems to be plenty of equipment – not all of it maxed out in capacity – so perhaps you really just need a system or software program that helps you maximize the capacity.

At a meeting of the American Mold Builders Association, one shop owner mentioned that he thought he needed more equipment until he put in a plant management system and found out that he was only utilizing 60% of his equipment time. That helped him see where he needed to maximize the capacity, and thus didn't have to add more equipment. You might consider something like that.

Name one short-term goal you would like to see the company accomplish in the next 90 days that would improve our relationships with current, satisfied customers.

- Replace some older equipment.
- A sonic cleaner to clean parts.
- Define targeted markets and players in those markets.
- Better marketing program to tell us how to break into new companies, and how to get in deeper at some of the companies we currently do business with.
- Diversify our customer base.
- Commitment to identify education needs and move forward with those.
- Compress timing - that will give us more opportunities if we can out-perform another shop.
- Adding another engineer.
- Increase engineering speed/shorten lead times.

Comments and suggested actions

Marketing and education seem to be the primary short-term goals. Adding another engineer and shortening lead times are also good goals that might take longer. There is the recognition that educational efforts are important in a mold shop setting, and can result in improved utilization and faster mold build times.

If we have lost business from current customers over the past year, why did we lose that work? How could we have kept that work (assuming it was business we wanted)?

Everyone who answered this question said that work is typically lost because of price, and this is the number one issue when a job is lost to a competitor. However "*When we sacrifice quality to meet a target price, the company's reputation suffers,*" said one respondent.

The second issue is lead time. "*We might still capture the work if we could deliver earlier.*" That is generally the case with the industry today. While many companies recognize quality, they will be willing to pay more if they can get the mold in a shorter time.

"*Communications. Knowing the issues and addressing the issues right from the start,*" was another comment. Communication is critical to the success of a program and to meeting lead times.

Part 1c: Potential customers/markets

On average how many RFQs do we receive monthly from current customers?

200 is the average.

On average how many RFQs do we received monthly from new, potential customers?

This was rather ambiguous – no one really knew but guesses ranged from 6 or 7 up to 15.

Comments and suggested actions

Most of the RFQs from current customers actually come from the same company, which I understand has no one to cost jobs at their facility so they send every requirement to ABC Moldmaking for quote. To have even 100 RFQs per month coming from one customer is extremely high, so 200 quotes from one customer is astronomical.

The fact that you have virtually no RFQs coming from new contacts means that few potential clients have ABC Moldmaking on a bid list. This really limits the company's ability to capture work from new companies or in new markets.

On average what percentage of RFQs that we quote turn into jobs?

XYZ Company – 1 in 30; John Doe Company – 1 in 5 ("*They are more specific in the types of things they have us quote,*" was one comment.)

Another person said 1 in 20-25.

½% to 1% of quotes turn into jobs.

"*XYZ represents 80% of the quotes we do but have a hit rate of 1% - 2%,*" was another comment.

Comments and suggested actions

This is extremely low. Most mold shops consider a 25% hit rate about average. Anything less than that means that you are quoting a lot of jobs that are not in your area of expertise. Several people also mentioned that ABC Moldmaking almost never 'no bids' a job.

You should be averaging 25% from current customers because they know you the best and know the types of molds that the company is best at designing and building. Hit rates for new customers will be lower because they do not know you well enough. Given the under staffing situation in engineering, I would recommend that the company be more discerning with the jobs quoted.

The situation with XYZ Company is particularly problematical. They appear to use a quote from ABC to assist them in costing every job, although very few of these jobs are then passed back to ABC. Perhaps you could agree to charge them for the quotes you provide for those molds that are outside your area of expertise. That way, you could still have an opportunity to quote the jobs you feel you really have a chance to get – and those are quoted at no charge – while helping out XYZ with their estimates for those molds that you won't get because of price, time, etc.

On the quotes we lost to a competitor, list the most common reason(s) given:

- Price
- Lead time

Comments and suggested actions

I think it's pretty obvious that we have to get more RFQs coming in from new, potential customers that are in ABC Moldmaking's area of expertise, and do less quoting for XYZ. The company has become quite dependent on XYZ but the hit rate is so small that it is apparent that there is a lot of effort going into XYZ with marginal reward.

In your experience, what is the most effective way to obtain new customers or enter new markets? (direct sales, advertising, better website, etc.)

- Direct sales person with proven contacts.
- Open the door then follow up with technical backup.
- Not just promoting the company but actually educating the customers.
- Buyers are buyers. We need a meeting to drill past the buyer – have to WOW somebody and get an advocate within the customer. Get the right people to come out.

- Get involved in product development stage, but we need an education component.

Comments and suggested actions

Most of the people interviewed were really at a loss as to how to do this, but had the right idea: we need to get in the door, find an advocate in engineering or product development, and educate the customer about why ABC Moldmaking is a valuable supplier. This is the key to developing long-term relationships with customers.

Rate the effectiveness of the following methods of promoting the company to gain new business (10 = most effective, 1 = least effective):

- Direct sales people (working only for our company) 3
- Independent sales reps 5
- Internet/website 3
- Print advertising (magazines, trade newsletters, etc.) 0
- Press releases
- Joining industry/trade associations, networking
- Word of mouth from current, satisfied customers

Comments and suggested actions

Very few of those interviewed had any idea about these things. Most said that a company has to have a good sales person, but doubted that the sales person without any technical knowledge could do any good.

No faith in the ability of independent sales reps, but actually ABC Moldmaking has had some success with sales reps. "They open doors and leave with something we can work with and leave an impression," was one comment.

Everyone appreciates that a website is essential today, but most people thought that ABC Moldmaking's site could be better: e.g. "It could be more educational, more case studies, why we choose the gates we do, white papers on molds", "We could do a better job of telling what we really do", and "We could put more photos on it".

What is one short-term goal (next 90 days) that you would like to see us achieve in the area of marketing and sales?

- We short-change ourselves in getting our name out – need to do that better.
- Get one new customer who is looking for a shop like us – not just a cheap mold builder.
- Create case studies and white papers – if you market yourself, you market your company.
- Marketing plan for the next year and 4 stories on ABC Moldmaking this year.

Part 2: Recommendations based on questionnaire returns in Part 1

Engineering/Project Management:

1. *Put in place an engineering manager who can oversee things in a 'big picture' way and get projects kicked off in a more timely way. Eric would seem to be a good choice for this – as a family member he has a big stake in the game and might have the ability to push things through. However, he will need to learn to temper his aggressiveness with some finesse since there are a couple of guys who are a bit sensitive to being liked.*

2. *Engineers/account managers need to have responsibility for certain accounts. Currently they feel 'pulled in too many different directions'. It appears that XYZ needs to have one engineer/account manager just for that account alone. Does Bill work well with XYZ? He's been at ABC Moldmaking the longest and given his need to be liked, it might be a good spot for him as XYZ is the biggest account. Tom seems to have a good temperament, even-keeled, knowledgeable, and seems comfortable and relaxed with people, which makes me think he can deal well with a variety of customers such as John Doe Company and the other companies that you only deal with periodically.*

3. *Tom has a real place in engineering and people who mention him say that he's very smart and creative, and comes up with good solutions to manufacturing problems when looking at new projects. He seems more quiet, the 'thinker' type, and good for working on projects with other engineers/account managers.*

4. *Engineering Coordinator – this might be a new role or an additional role for someone like Susie who currently covers customer service. An engineering coordinator logs in RFQs, prepares the quotes for Engineering, logs the out-going quotes, tracks the quotes, and logs the quotes as jobs when the P.O. comes in.*

Customers should come to use her as a point of contact for getting questions answered if the engineers are difficult to reach or don't have time to do a lot of calling. Would take some of the 'busy work' away from the engineers and give them more 'project management' time.

5. *The number of quotes from XYZ is really amazingly high. Given the low hit rate, it might be better to superficially review the RFQs, then quote only those jobs you are certain that you will get. Throwing numbers on all of the RFQs is okay if you have the time, but if engineering is spread thin your time might be better spent really trying to nail the jobs you know you've got a decent opportunity to land. Average hit rate for a shop that really knows its business is about 25% for current, satisfied customers. It's less – probably 1 in 20 – for new, potential customers who are just feeling you out for the first time. That rate should improve as you become better acquainted with the company and they become better acquainted with you and your capabilities. Closely evaluate the cost of doing business with XYZ. Do they take so much time that you have little spare to focus on getting new work from new customers? Do they really pay for what they consume of ABC Moldmaking?*

6. *Focus has got to be the name of the game, and you have to remember where your value is when looking at the jobs you are asked to quote. This will be especially true as you push your efforts into new markets, new customers, etc.*

Sales:

7. *Bill is an outgoing, enthusiastic person who loves sales and that's good. However, I think he lacks confidence because he feels he doesn't know about mold building. He needs to be encouraged to do the sales job and not worry about his lack of technical expertise. He seems to have the backup from engineering or Bob if he needs it, but I think he just feels too intimidated by 'engineering' types when he's cold-calling.*

Systems:

8. *A good plant management software system might be helpful. This would help you track your jobs more closely, give you a better look at machine utilization to see where you have bottlenecks and where you have under-utilized machinery. There are several on the market that you might look at. You could also provide customer progress reports online using customer-specific secure locations on your website.*

New business:

9. *Several people mentioned the repair department operating as a separate profit center. This might be a good thing, as you can offer your repair services to anyone in the area who needs quick-turn on a mold repair. It would appear*

that you have the skill and good equipment to offer this. I know one shop that mans its repair facility around the clock, with the offer that anytime a mold goes down and needs repair, they've got a repair person on duty. It's actually become a good source of revenue for this shop, which is quite a bit smaller than ABC Moldmaking.

10. *Everyone is very aware that you have a very narrow customer base, and I think there is a sense that something needs to be done about getting more customers. They almost have a sense of worry about this lack of customers, and the feeling that you are so locked into XYZ that their business is more of a burden than a blessing. This is not such a bad situation. People see the need to promote the company and look for new customers and jobs, whilst also appreciating that there is a relatively secure supply of work while these new contacts are still coming through.*

Defining and understanding your value:

11. *ABC Moldmaking brings a lot of value to the table with an extensive array of capabilities in machine tool technology, moldmakers and machinists, engineering talent, and creative and innovative people who can come up with unique manufacturing solutions for customers, but this is undersold.*

12. *Think bigger than the automotive industry:*

"ABC is a mold engineering and manufacturing firm with extensive capabilities, providing innovative thinking and creative solutions to mold manufacturing challenges."

13. *Develop a value proposition that everyone can buy into, such as the following:*

"ABC Moldmaking is a mold engineering and manufacturing firm, specializing in the design and development of challenging, mechanically-complex molds, utilizing state-of-the-art machine tools and automation technology, employing experienced moldmakers dedicated to providing the best manufacturing solutions for ABC Moldmaking's customers."

14. *Promote your value through case studies and white papers on your website; through promotional literature sent via mail, and by educating your customers in a number of ways:*

- Customer appreciation days
- Education seminars
- One-on-one meetings with customers and potential customers to explain your value

15. *Match your customers' and potential customers' requirements with your capabilities.*

Part 3. Information gathering, round tables and SWOT evaluations

The SWOT evaluation process is designed to help you match your capabilities and expertise with those of companies in various industries that require what ABC Moldmaking has to offer.

SWOT evaluations are frequently carried out in the format of a round table discussion, where everybody's opinions can be considered. However, this process needs to be well planned and managed in order to derive the best results. Where you are evaluating those external aspects which influence your strategy it is important that individuals are tasked with gathering information in advance, to avoid wasting time on speculation. The actual process can be carried out using a whiteboard divided into four sections, with the various facets of the company or sector under discussion being placed in the most appropriate 'quarter'.

When carrying out a SWOT evaluation, you are attempting to define the Strengths, Weaknesses, Opportunities and Threats associated with a particular business or market. If you go through this process with all aspects of a company and the business environment in which it operates, you can then go on to match up your strengths with the best opportunities, or those areas where you customers or competitors are weakest.

At the stage which ABC Moldmaking has reached above, they already know a great deal about themselves. It should now be simple to sit down and work out how this information feeds into a SWOT chart, whilst getting people into the right way of thinking for subsequently applying the same principles to external factors. Once they have completed this self-evaluation they can then move on to the world outside their four walls. At this stage you move away from factors which you can affect directly, but start to work out where best to direct your efforts.

SWOT evaluation of the moldmaking industry

This can be a fairly broad-reaching discussion, as many people around the table will have experience of the wider industry through reading the press, previous job experience, or attending trade shows. Strengths and weaknesses may be related to more general economic conditions or broad changes in plastics manufacturing or applications. Opportunities could be provided by new moldmaking technology or ways to be more competitive. ABC should be looking at high-speed machining and ways of improving delivery times. Threats could include offshoring of moldmaking or molding customers.

SWOT evaluation of the end-user industries

This stage of the process can require some research, but again it is wise to start with what you know, and then apply the same approach to less familiar subjects. So ABC Moldmaking would start by carrying out a SWOT analysis of the automotive molding sector, where for example strengths could include the trend towards weight reduction for energy efficiency, favouring the use of more plastics components. The auto industry also thrives on regular model changes, which can be seen as an opportunity for molders and their suppliers to feed a regular demand for new car body parts.

Having once got a feel for a familiar sector, they can then consider other potential sectors. This is where an individual might be tasked with some advance research, to look at technological advances or other changes in the business environment. Important factors to consider when looking at a new market sector may be as diverse as the overall size of the market for plastics moldings, geographical location of the market, typical products (how do they compare with your current output), regulatory or environmental pressures influencing demand, and even demographics affecting markets such as medical and health-care products. You will probably not get a really deep understanding of a new market sector at this stage, but you can start to see where your capabilities match particular businesses.

SWOT evaluation of current and potential customers

This stage of the process is less suited to a round table discussion and may be best carried out by those individuals with most customer contact using a standard template to allow comparisons. The aim of the game is to look at current customers and determine their value to you in terms of current profitability and potential growth. The information about current customers will come from the customer questionnaires and also records of past business with that customer. ABC obviously need to look really closely at the relationship with their biggest customer XYZ. Once you have worked out what constitutes your 'ideal customer', you can also work out what criteria to apply when choosing potential new target customers. You may be looking particularly for customer weaknesses here; for example your ideal customer may have difficulty doing short-run molding trials or new product testing, whereas these could be good areas for you to add value.

Analysing your competition

This is the final part of the information gathering stage, and the initial data comes from staff and customer surveys and broader industry intelligence. When you have identified your competitors, look at what they have to offer in terms of technology, service and added value (all this is easily found on websites). In the case of ABC Moldmaking's competitors, none of them were any more outstanding than ABC. All offered similar machine capabilities, engineering expertise, and many of the same services. Most promoted themselves in terms of their machinery and equipment, moldmaking capabilities and molding facilities.

Part 4. Putting it all together

So far you will have been through a huge information gathering and learning exercise, but this has the potential to be largely wasted effort unless you then use your accumulated knowledge. Now you need to take the key facts and build them into a strategy.

So start again with what you know about your own company – where do you need to invest time and money? Improving the offer or finding new customers? It's probably a combination of both.

Then look outwards. Are your current markets thriving, changing or just declining? Do you want more of the same customers or new ones with different needs but potentially better rewards? ABC certainly need to look outside the difficult automotive business, but stay in relatively high-tech, high-value areas that appreciate their expertise.

Build a strategy for the next twelve months or longer, with a variety of short, medium and longer term goals and share it with everyone involved. Having been through the development process they should be as keen as you to share your vision and see it work. Short term goals could include improving the current offer (customer service, project management, training ...) investing in urgently needed equipment, software or repairs, and working on the image you present through your website. Medium and longer-term goals may involve greater investment in equipment or in the time needed to locate and cultivate new customers and markets. Set targets, deadlines and ways of measuring your success. And finally, make sure that you use this strategy in managing the direction of your company. Refer to it frequently, evaluate which parts are working best for you, and modify your targets if circumstances change.

Glossary of abbreviations

ABS	Acrylonitrile-butadiene-styrene terpolymer
AMBA	American Mold Builders Association
APR	Association of Postconsumer Plastic Recyclers
B2B	Business to business
BCC	Best-cost country (for manufacturing / outsourcing)
BOM	Bill of materials
CAD	Computer aided design
CAGR	Compound annual growth rate
CD	Compact disc
CEMC	Clean environment manufacturing cell
CNC	Computer numerical control
COC	Cyclic olefin copolymer
COD	Cash on delivery
CPA	Certified public accountant
CRM	Customer relationship management
DMLS	Direct metal laser sintering
DPI	Dots per (square) inch
ECU	Electronic control unit (vehicle)
EDM	Electrical discharge machining
ELV	End-of-life vehicle (European Commission directive 2000/53/EC)
EOAT	End-of-arm tooling
EPA	Environmental Protection Agency (United States)
ERP	Enterprise resource planning
ESOP	Employee stock ownership plan

EuPC	European Plastics Converters (trade association)
EV	Electric vehicle
FDM	Fused deposition modeling
FIM	Film insert molding
FTP	File transfer protocol
GMP	Good manufacturing practice
HDPE	High density polyethylene
HEPA	High efficiency particulate arresting or high efficiency particulate air (filters)
HR	Human resources
HSM	High speed machining
IMD	In-mold decorating
IML	In-mold labeling
IP	Intellectual property
JIT	Just in time
JV	Joint venture
LCC	Low-cost country (for manufacturing / outsourcing)
LCP	Liquid crystal polymer
LFRT	Long fibre reinforced thermoplastic
LSR	Liquid silicone rubber
M&A	Mergers and acquisitions
MEMS	Micro-electro-mechanical system
MRP	Material requirement planning
MUD	Multi-unit die (base)
NAPCOR	National Association for PET Container Resources
OEE	Overall equipment effectiveness
OEM	Original equipment manufacturer
P&L	Profit and loss
PA	Polyamide (nylon)
PBT	Polybutylene terephthalate (polyester)
PCC	Post-consumer content
PCR	Post-consumer recyclate
PE	Polyethylene
PEEK	Poyetheretherketone
PERC	Programmable Electric Rotating Cores (BA Die Mold system)

PHA	Polyhydroxy-alkanoate
PHB	Polyhydroxybutyrate
PHBV	Polyhydroxybutyrate-valerate
PLA	Polylactic acid
PM	Preventive maintenance
PMS	Pantone (colour) matching system
PO	Purchase order
POM	Polyacetal, also called polyoxymethylene
PP	Polypropylene
PPAP	Production part approval process
PPS	Polyphenylene sulphide
PR	Public relations
PS	Pressure sensitive
REACH	Registration, Evaluation, Authorisation and Restriction of Chemical substances (European Commission regulation 1907/2006)
RFQ	Request for quotation
RoHS	Restriction of the use of certain Hazardous Substances (European Commission directive 2002/95/EC)
ROI	Return on investment
RPET	Recycled polyethylene terephthalate
SFDA	State Food and Drug Administration (China)
SLA	Stereolithography
SLS	Selective laser sintering
SSD	Solid state drive
TPO	Thermoplastic polyolefin elastomer
TRIM	Thin recessed injection molding (StackTeck patented technology)
WEEE	Waste Electrical and Electronic Equipment (European Commission directive 2002/96/EC)

Index

3D printing 204
  multi-material 206

A

accounting
  software 154
acquisitions 77, 84, 90
  bankrupt companies 84
  secondary operations 84
additive manufacturing 203
  business models 205
  low cost 206
  low-volume 210
  micro-electro mechanical exposure 206
  personalization 205
  service bureaus 212
  surgical instruments 211
  thermoplastic materials 207
advanced technologies 195
aerospace products 117
  certifications 117
air leaks 188
appliances 89, 118
automotive molding 2, 5, 85, 99, 101, 123, 192, 218
  cost reductions 104
  weight reduction 104

B

bankrupt companies 84
banks 79
best-cost countries 138
biodegradable plastics 222
bioplastics 222
branding 42
building applications 122
  roofing products 122
business cycles 76
business operations 149
  partnerships 233
  software 153
  stock ownership 234
business plans 76, 79
  business valuation 229
  executive summaries 81
  finances 83
  inventories 82
  long term 229
  marketing strategies 83
  markets 82
  objectives 81
  organisation chart 82
  sales strategies 83
  selling up 229
  skills audits 83
  software 83
  staffing 82
  timescales 81
  vision 79, 81

C

carbon footprint 141
case histories 45, 72
  additive manufacturing 210
  advanced technologies 195
  aerospace products 118
  automotive materials 104
  back-shoring 136
  clean environment cells 184
  diversification 85
  early growth 1
  lead times 192
  lean manufacturing 143
  marketing communications 45
  medical molding 4
  mold transfers 72
  packaging 114
  plant modernization 146
  precision molding 129
  rapid prototyping 213
  specialization 129

stock ownership 234
case study 239
cash flow 101, 145
certification 129
clean rooms 27, 183
co-injection molding 140, 168
communications 40
advertising 55, 78
internal 74, 90
internet marketing 50
networking 50,85
online news services 55
press coverage 51
websites 46
competitiveness 138
moldmaking 192, 198
competitors 31
compounds 219
computer applications 124
conferences 55
consolidation 89, 101
appliances 118
packaging 111
construction applications 122
costing 216
inventory 156
moldmaking 135, 196
sales 156
software 154
customer relationship management
software 154
customer surveys 25, 38
problem customers 69
questionnaires 23
cycle times
robotics 180

D

decoration 182
film insert molding 100
metalizing 104
defense products 117
certifications 117
digital prototyping 209
direct metal laser sintering 204, 209, 211
diversification 69, 85, 151, 152
in-mold labeling 113

E

electrical components 123, 219
electrical products 123, 175
electric machines 178
unscrewing molds 195
electronic components 123, 219
automotive applications 123
computer applications 124
energy efficiency 141, 142, 145
ancillary equipment 145
lighting 145,146
enterprise resource planning 155
software 153
expansions 84
external gas molding 175

F

factory organization 178
film insert molding 100, 170
financial management 83
fluid assisted injection molding 175
foam molding
microcellular 174
structural 173
fuel resistance 218
fused deposition modeling 203, 210

G

gas assisted injection molding 175
globalization 9, 89, 131
appliances 119
automotive industry 102
electronic equipment 123
medical molding 108, 109
moldmaking 131, 191
packaging 115
toys 120
green materials 221

H

high performance polymers 100, 104, 121, 124, 125, 218
electronic components 220
industrial products 117
micromolding 175
high-speed machining 194

history 1
additive manufacturing 203
automation 3
compact discs 6
consumer electronics 6
decorating 3
early days 1
electric machines 3
globalization 9
medical molding 4
mobile phones 6
moldmaking 10
packaging 1
personal computers 4
rapid manufacturing 203
recession 11
secondary operations 3
hot stamping 182

I

IMMEX operations 133
legislation 134
industrial products 116
industry guides 32
industry size 12
information resources
industry guides 12, 32
market research 99
statistics 12
trade associations 99
in-mold decorating 168
in-mold labeling 112, 157, 168
costing 112
market research 112
innovations
packaging 111, 114
insert molding 118, 170
intellectual property 110, 136, 139, 230
internet marketing 50
inventory 82, 145, 217
cash flow 145
costing 156
software 154

L

lay-offs 89, 91
lean manufacturing 143
leasing 78
long term planning 229
low-cost countries 138
low pressure injection molding 173

M

machine utilization 150, 153
maintenance 185
air leaks 188
barrels 189
energy efficiency 186
filters 187
lubrication 187
scheduling 188
manufacturing cells 179
clean room 183
moldmaking 194
maquiladora operations 133
legislation 134
margins 157
marketing 17
branding 42
differentiation 34, 43
testimonials 35
marketing strategies 18, 19
self examination 19
market research 12, 99, 126
bioplastics 222
in-mold labeling 112
sales growth 150, 151
sports equipment 120
material costs 216
material requirements planning 154
medical molding 4, 27, 105
addtive manufacturing 210
approvals 106
clean room 183
good manufacturing practices 107, 108
intellectual property 110
market size 105
offshoring 136
precision molding 129
protocols 107
regulations 108
microcellular foam molding 174
micromolding 175
electronic components 175
measurements 176
mechanical properties 176
medical molding 176

mobile phones 175
military products 117
certifications 117
mobile phones
film inset molding 170
mold components 199
moldmaking 191
3D design 193
CAD/CAM 192
components 201
direct metal laser sintering 209
high-speed machining 194, 200
lead times 192, 200
mold bases 199
offshore 103, 135, 196
precision 195
prototyping 213
unscrewing molds 195
mold transfers 71
motorcycles 121
MuCell 174
multi-material molding 166

N

networking 85

O

offshoring
back-shoring 141
medical molding 136
moldmaking 191, 196
ordering
software 154
overall equipment effectiveness 185
over-capacity 153
overmolding 166

P

packaging 1, 111
good manufacturing practices 116
in-mold labelling 112
innovations 111, 114
personal care 114
transit 173
pad printing 183
painting
elimination 167
personal care products 126, 221
plant-based plastics 222
precision molding
electronic connectors 123
press coverage 51
press releases 52
trade magazines 53
pricing 28, 152
margins 156
problem customers 69, 78, 158
production monitoring
software 154
productivity 89
stack molding 171
profitability 149
checklist 161
machine utilization 150
problem customers 158
value add 150
volume 150
proprietary products 7, 84, 129, 130
packaging 113
purchasing 78, 216
purging 147

Q

quality control 137
questionnaires 21, 23, 36
customer surveys 38
self examination 36
quoting 28, 30, 74
software 154

R

rapid manufacturing 203
rapid prototyping 203, 208
plaster-based 207
service bureaus 203
raw materials 216
recessions 76, 101, 191
recovery 151
recyclability
packaging 114
recycling 140, 142
bioplastics 224
co-injection molding 168
waste stream contamination 224
regrinding 147

retirement 229
robotics 179
in-mold labeling 113, 169
rotational stack molding 172

S

sales 58
costing 66, 156
direct sales 60
inside sales 61
sales personality 59, 62
sales reps 60
sales staff 58
sales strategies 63
targeted marketing 65
sales strategies 58
mold transfers 71
new customers 70
priority customers 68
target customers 66
sandwich molding 168
scheduling
software 154
scientific molding 46
secondary operations 78, 181
additive manufacturing 212
selective laser sintering 203
self examination 19, 36
questionnaires 21
surveying customers 23
surveying employees 21
short run molding 45
skills audits 83, 91
skills matrix 93
software 153
CAD/CAM 192
maintenance 188
specialization 128, 129
precision molding 129
sports equipment 120
in-mold labeling 169
market size 120
stack molding 171
rotational 172
staffing 82, 86
statistics 12
stereolithography 203
structural foam molding 173
large components 173
super light injection molding 174
supply chain flexibility 136
sustainability 140
carbon footprint 141
energy efficiency 141, 142, 145
lean manufacturing 143
maintenance 185
purging 147
recycling 142
regrind 147
transit packaging 146
waste reduction 142, 147
SWOT analysis 18, 24, 27, 78

T

tandem molding 171
target customers 66, 78
templates 92
customer surveys 38
self examination 36
training plans 92
thermoplastic elastomers 105, 221
thin recess injection molding 157, 169
toxic employees 90
toys 120
trade magazines 53
trade shows 55
training 86
benefits 88
cross-training 87
plans 91
return on investment 88
two-shot molding 165
hard/soft combinations 167
reaction injection molding 167

U

used equipment 76, 179

V

value model 29
value proposition 17, 27
virtual prototyping 209
vision 79, 81

W

waste bins 140

waste reduction 142, 147, 217
water assisted injection molding 175
websites 46
  content 47
welding 181
work cells 144

FSC
www.fsc.org
MIX
Papier | Fördert gute Waldnutzung
FSC® C083411

Kolibri 360 GmbH
Ferdinand-Jühlke-Straße 7
99095 Erfurt, Deutschland
produktsicherheit@kolibri360.de